Springer Tracts in Modern Physics
Volume 154

Managing Editor: G. Höhler, Karlsruhe

Editors: J. Kühn, Karlsruhe
Th. Müller, Karlsruhe
R. D. Peccei, Los Angeles
F. Steiner, Ulm
J. Trümper, Garching
P. Wölfle, Karlsruhe

Honorary Editor: E. A. Niekisch, Jülich

Springer
Berlin
Heidelberg
New York
Barcelona
Hong Kong
London
Milan
Paris
Singapore
Tokyo

Springer Tracts in Modern Physics

Springer Tracts in Modern Physics provides comprehensive and critical reviews of topics of current interest in physics. The following fields are emphasized: elementary particle physics, solid-state physics, complex systems, and fundamental astrophysics.

Suitable reviews of other fields can also be accepted. The editors encourage prospective authors to correspond with them in advance of submitting an article. For reviews of topics belonging to the above mentioned fields, they should address the responsible editor, otherwise the managing editor. See also http://www.springer.de/phys/books/stmp.html

Managing Editor

Gerhard Höhler

Institut für Theoretische Teilchenphysik
Universität Karlsruhe
Postfach 69 80
D-76128 Karlsruhe, Germany
Phone: +49 (7 21) 6 08 33 75
Fax: +49 (7 21) 37 07 26
Email: gerhard.hoehler@physik.uni-karlsruhe.de
http://www-ttp.physik.uni-karlsruhe.de/

Elementary Particle Physics, Editors

Johann H. Kühn

Institut für Theoretische Teilchenphysik
Universität Karlsruhe
Postfach 69 80
D-76128 Karlsruhe, Germany
Phone: +49 (7 21) 6 08 33 72
Fax: +49 (7 21) 37 07 26
Email: johann.kuehn@physik.uni-karlsruhe.de
http://www-ttp.physik.uni-karlsruhe.de/~jk

Thomas Müller

Institut für Experimentelle Kernphysik
Fakultät für Physik
Universität Karlsruhe
Postfach 69 80
D-76128 Karlsruhe, Germany
Phone: +49 (7 21) 6 08 35 24
Fax: +49 (7 21) 6 07 26 21
Email: thomas.muller@physik.uni-karlsruhe.de
http://www-ekp.physik.uni-karlsruhe.de

Roberto Peccei

Department of Physics
University of California, Los Angeles
405 Hilgard Avenue
Los Angeles, CA 90024-1547, USA
Phone: +1 310 825 1042
Fax: +1 310 825 9368
Email: peccei@physics.ucla.edu
http://www.physics.ucla.edu/faculty/ladder/
peccei.html

Solid-State Physics, Editor

Peter Wölfle

Institut für Theorie der Kondensierten Materie
Universität Karlsruhe
Postfach 69 80
D-76128 Karlsruhe, Germany
Phone: +49 (7 21) 6 08 35 90
Fax: +49 (7 21) 69 81 50
Email: woelfle@tkm.physik.uni-karlsruhe.de
http://www-tkm.physik.uni-karlsruhe.de

Complex Systems, Editor

Frank Steiner

Abteilung Theoretische Physik
Universität Ulm
Albert-Einstein-Allee 11
D-89069 Ulm, Germany
Phone: +49 (7 31) 5 02 29 10
Fax: +49 (7 31) 5 02 29 24
Email: steiner@physik.uni-ulm.de
http://www.physik.uni-ulm.de/theo/theophys.html

Fundamental Astrophysics, Editor

Joachim Trümper

Max-Planck-Institut für Extraterrestrische Physik
Postfach 16 03
D-85740 Garching, Germany
Phone: +49 (89) 32 99 35 59
Fax: +49 (89) 32 99 35 69
Email: jtrumper@mpe-garching.mpg.de
http://www.mpe-garching.mpg.de/index.html

Colin Stanley

August 2008.

Wolfgang Braun

Applied RHEED

Reflection High-Energy Electron Diffraction During Crystal Growth

With 150 Figures and 7 Color Plates

 Springer

Dr. Wolfgang Braun

Paul-Drude-Institut für Festkörperelektronik
Hausvogteiplatz 5–7
D-10117 Berlin, Germany
Email: braun@pdi-berlin.de

Physics and Astronomy Classification Scheme (PACS): 61.14Hg, 61.14Dc, 81.15Hi

ISSN 0081-3869
ISBN 3-540-65199-3 Springer-Verlag Berlin Heidelberg New York

Library of Congress Cataloging-in-Publication Data applied for.

Die Deutsche Bibliothek – CIP Einheitsaufnahme

Braun, Wolfgang: Applied RHEED: reflection high-energy electron diffraction during crystal growth/Wolfgang Braun. – Berlin; Heidelberg; New York; Barcelona; Hong Kong; London; Milan; Paris; Singapore; Tokyo: Springer, 1999 (Springer tracts in modern physics; Vol. 154)
ISBN 3-540-65199-3

© Springer-Verlag Berlin Heidelberg 1999
Printed in Germany

Typesetting: Camera-ready copy by the author using a Springer TEX macro package
Cover design: *design & production* GmbH, Heidelberg
Computer-to-plate and printing: Mercedesdruck, Berlin
Binding: Buchbinderei Lüderitz & Bauer, Berlin

SPIN: 10674314   56/3144 - 5 4 3 2 1 0 – Printed on acid-free paper

Preface

The precisely controlled preparation of samples is of paramount importance in both solid-state physics and materials science. In addition to its success as a device production technology, molecular-beam epitaxy (MBE) is probably the most important method in this field, since it allows the growth of crystalline layer combinations with accurate dimensional control down to the atomic level. This precision would not be possible without adequately accurate characterization techniques like reflection high-energy electron diffraction (RHEED) that provide resolution on the atomic scale while at the same time being fully compatible with the crystal growth process.

The increased understanding of RHEED in the last three decades is inseparably connected with the vigorous development of MBE during this period. Both techniques have profited from each other, each new achievement on one side triggering new developments and exciting breakthroughs on the other. The most notable example is the discovery of RHEED intensity oscillations in 1980, which was made possible by the increased purity of MBE source materials as well as improvements in the achievable vacuum levels. RHEED intensity oscillations in turn greatly facilitated MBE growth as a convenient way to calibrate growth rates. RHEED nowadays is an almost indispensable tool for the MBE crystal grower. The body of literature on RHEED is substantial, with several theoretical as well as experimental groups having been dedicated to the subject for years. Theoretical treatments have reached an advanced degree of sophistication, and many aspects of RHEED have been studied extensively and in great detail.

On the other hand, the dynamical (multiple-scattering) nature of RHEED complicates the interpretation of the data. Many scientists and engineers approaching the subject from the perspective of MBE growth therefore find it difficult to obtain meaningful growth parameters from their RHEED data, and RHEED still has a somewhat mysterious aura.

This book attempts to bridge this gap by providing an introduction to RHEED for scientists and engineers who not only want to study RHEED for its own sake, but need to apply it as a tool to study surfaces and interfaces. At the same time, I intend to demonstrate how straightforward experiments can provide insight into the fundamentals of RHEED without having to resort to very involved theoretical models. It is my intention and hope that this book may be useful to crystal growers, surface scientists and dedicated RHEED researchers alike.

Many people have contributed to this work, which summarizes several years of work at the Max-Planck Institute for Solid State Research in Stuttgart, the Paul-Drude Institute for Solid State Electronics in Berlin and Arizona State University in Tempe, Arizona. It would not have been possible without the constant support and encouragement of K. H. Ploog, who accompanied and nurtured it from the very start. Many discussions with L. Däweritz initiated and shaped several parts, especially the semikinematical simulations of Chap. 4. Y.-H. Zhang supported the work on the main theoretical conclusions and the chapter on RHEED with rotating substrates. A close collaboration with M. Wassermeier and J. Behrend (STM), and A. Trampert (TEM), who contributed several figures, helped to provide more detailed insight into structural aspects of growing surfaces and segregation mechanisms.

A. Fischer introduced me to the secrets of MBE growing (and its history) and taught me all of the technological details necessary to keep a good MBE system up and running. His decades of experience form the engineering foundation of this book.

I also want to thank C. Irslinger at the Max-Planck Institute for his constant support throughout my work there. The librarians, computer experts, engineers, technicians, photographers and machine shop people at the various institutions contributed a not directly visible, but significant part to this work.

Special thanks go to F. Große, K. Hagenstein, M. Höricke, H. Möller, J. Müllhäuser, M. Reiche, O. Brandt, P. Dowd, P. Schützendübe and D. K. deVries for various contributions. This work would not have been possible without the constant support and encouragement of my family, especially my wife, who bore with me in the tedious times.

Last but not least, I would like to thank P. Wölfle, H. J. Kölsch, J. Lenz, V. Wicks, P. Treiber, U. Heuser and the production staff at Springer for their encouragement and support in the production of this monograph.

Berlin, January 1999   *Wolfgang Braun*

Contents

1. MBE-Grown Semiconductor Interfaces

Reflection high-energy electron diffraction (RHEED) as a characterization technique cannot be demonstrated without applying it to a particular surface. The model systems used throughout this book will be the GaAs/AlAs (001) and (113)A surfaces grown by molecular-beam epitaxy (MBE). We therefore start with a brief introduction to MBE and an introductory review of the surfaces and interfaces that will be studied. Most measurement methods, however, are generally applicable to many materials systems and deposition methods, as demonstrated by the large number of publications in the field. The choice of a common model system merely serves the purpose of focusing on the method instead of the material.

1.1 Molecular-Beam Epitaxy

Since its conception in the 1960s [1,2], molecular-beam epitaxy has experienced a tremendous development. Taking the scope of the 1996 MBE conference in Malibu as a benchmark, the field of MBE nowadays includes the growth of group IV, III–V and II–VI semiconductors, metals, magnetic materials, nitrides, oxides and fluorides using solid and gaseous as well as metalorganic sources. For the study of interfaces, including special systems such as polar on nonpolar materials and strained heteroepitaxial systems, as well as fundamental studies of dopant incorporation and activation, MBE still stays at the forefront of basic research. This success of MBE is due to the versatility and simplicity of its operating principle: in an ultrahigh vacuum (UHV) environment, beams of atoms or molecules are directed onto a heated substrate crystal, where they form a crystalline layer. In Fig. 1.1, a cross-section of a basic MBE system for solid source materials is represented.

Partly because of the technological difficulties associated with UHV, molecular beam epitaxy has had a slow start in device production. Many epitaxial devices are still fabricated by metal-organic vapor phase epitaxy (MOVPE). The need to use MBE for some devices like high-mobility modulation-doped heterostructures, where it is the only possible fabrication method, has spurred the development of high-throughput production MBE systems that are now cost-effective enough to compete with traditional growth methods for almost any device. One of MBE's big advantages in this context is the

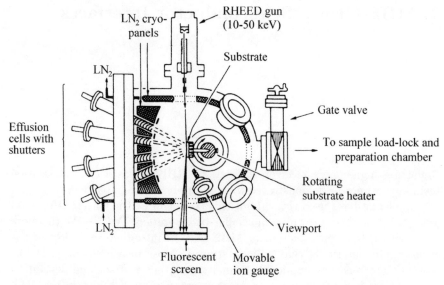

Fig. 1.1. Top view of a typical standard MBE system growth chamber [3]

fact that hazardous chemicals are usually handled in solid form and contained within the vacuum vessel, reducing the cost of external safety measures. We can summarize four major advantages that make MBE a supreme tool for basic research as well as device production:

- MBE allows a very precise control of layer thickness and dopant incorporation down to the atomic scale.
- Compared to other growth techniques, no complicated chemical reactions take place at the substrate surface. This facilitates analysis of growth processes such as surface migration and dopant incorporation.
- The UHV environment in the growth chamber allows the application of various in-situ measurement techniques to study the processes governing crystal growth. At the same time, these measurements can be used to implement real-time feedback loops for growth control.
- Toxic chemicals are contained within the vacuum chamber.

The main components of an MBE system are shown in Fig. 1.1. The source materials, in elemental form, are evaporated from the liquid or sublimed from the solid phase. For the studies described here, As, Ga, Al, Si, Sn and C were used to produce doped and undoped $Al_x Ga_{1-x} As$ ($0 \leq x \leq 1$) layers. Except for C, which is sublimed from a resistively heated filament, all source materials are contained in pyrolytic boron nitride crucibles to reduce impurity contamination. The sources are electrically heated with a temperature stability of typically $0.1\,°C$. The molecular and atomic beams can be switched on and off by shutters in front of the crucibles, which are actuated by computer-controlled motors. The single-crystal GaAs substrate is attached to a heatable manipulator that allows continuous rotation of the sample to improve uniformity.

Generally, determination of the sample temperature is difficult. The thermocouple, which is placed near the back of the sample holder, has a large temperature offset with respect to the sample surface and is only weakly coupled to the radiatively heated substrate [4]. The oxide desorption temperature of GaAs ($\approx 580\,^\circ$C) can be used for calibration, but is not very accurate as it depends on the As$_4$ flux and sample preparation [5]. Some progress in the measurement of exact surface temperatures has been made recently by pyrometric interferometry [6] and band-gap thermometry [7], but accurate substrate temperature measurement during MBE growth remains a problem.

Fabrication of high-quality materials by MBE requires very low partial pressures of contaminants such as oxygen and hydrocarbons (below 1×10^{-10} Pa). Therefore, special materials need to be used for hot parts inside the vacuum. In addition, the complete system needs to be designed to withstand $200\,^\circ$C baking prior to crystal growth. Pumping is achieved with a combination of cryopumps, ion pumps, Ti sublimation pumps and several liquid-nitrogen-cooled shrouds that cover most of the inside chamber surface. The solid angle in front of the substrate is occupied by the sources in an MBE chamber. Apart from a small window that frequently allows optical measurements at normal incidence to the sample, this restricts access for measurement to lateral incidence and exit if no source ports are to be sacrificed for the purpose. Therefore, RHEED has become the most widely used analytical tool in MBE among the variety of methods implemented [8].

1.2 Interface Formation

Since the very precise control of layer thickness is one of the main advantages of MBE, the study of interface structures and their influence on device parameters has stayed in the mainstream of MBE research since the initial proposal of semiconductor superlattices and quantum wells by Esaki and Tsu [9, 10] in the early 1970s. In particular, the interfacial atomic configuration of GaAs/Al$_x$Ga$_{1-x}$As ($0 \le x \le 1$) heterostructures has played a major role in the development of modern semiconductor physics [11] and in the progress of advanced device concepts [12], since this materials system is commonly used as a model system for fundamental studies. It was not until 1991 that coherent tunneling, where the phase of the wave function is not changed by elastic scattering processes, was observed in high-quality samples [13]. Recently, the quantum cascade laser, a fundamentally new solid-state device based on the heterostructure concept, was presented [14]. At the same time, cleaved-edge overgrowth [15,16] and the use of self-organization mechanisms during MBE growth [17,18] are opening up new ways to further reduce the dimensionality of semiconductor structures. All these approaches rely critically on the interface quality, which is the most important parameter for defining quantum mechanical confinement.

During the development and evolution of MBE, the continuous refinement of growth methods and the appearance of more sophisticated characterization tools has led to an improved understanding of the complicated interplay between growth conditions and interface structure. The assumption that MBE produces atomically smooth and abrupt interfaces on a large scale had to be abandoned in this process. Instead, a variety of competing phenomena were found to be involved in the process of interface formation leading to various degrees of perfection. To focus on the main effects, we restrict the following summary to the (001) surface and to the growth of $Al_xGa_{1-x}As$ heterointerfaces with $0 \leq x \leq 1$.

A simple way to assess the interface structure consists of probing carrier confinement in quantum wells (QWs) using photoluminescence (PL) and photoluminescence excitation (PLE) spectroscopy. The excitonic luminescence, which dominates the optical spectra of many III–V compounds, is very sensitive to the shape of the barriers defining the quantum well. For some continuously grown quantum wells, a single peak is seen in PL or PLE, whereas other QWs prepared with growth interruption show a splitting of the PL/PLE lines, with negligible variation across the sample. This splitting is attributed to discrete thickness variations of the QWs on a length scale larger than the exciton diameter. A thickness variation smaller than the exciton diameter results in broadened peaks, observed for samples grown without growth interruption. For this small-size roughness, the wave function of the exciton averages over the fluctuations [19, 20].

Scanning near-field optical microscopy (SNOM) experiments [21] allow us to directly image the real-space morphology of the areas of constant thickness. Depending on the growth conditions, different roughness distributions are detected. Generally, samples grown around 580 °C with growth interruption at the interfaces show large areas of constant thickness, whereas samples prepared without growth interruption at higher temperatures exhibit higher spatial frequencies. Areas from micron size to the resolution limit of SNOM of about 0.1 μm can be directly imaged, whereas roughness on a smaller scale down to the confinement size of the excitons of ~ 50 Å is inferred from the spectral distribution of the individual quantum eigenstates. Theoretical models to analyze the relative peak positions of the split peaks are not yet accurate enough to extract more detailed information about the interface morphology. The split peaks are not equally spaced and show fine structure due to the energy difference between free and bound excitons [22]. An analysis of the split-peak spacing in terms of a roughness spectrum [23] therefore seems rather speculative at present. The best fit for optical absorption data is achieved if an asymmetry in the confining quantum well is assumed, which can best be explained by segregation at one or both of the interfaces [24, 25].

The same is true for Raman spectra, where a good fit between experiments and theory can only be achieved if nonabrupt interfaces are assumed [26,27].

High-resolution transmission electron microscopy (HRTEM) allows the real-space imaging of semiconductor crystals with atomic resolution. The resulting image is obtained from a slab typically 50 to 100 lattice constants thick. Therefore, a unit cell in an HRTEM picture represents information averaged over a column of unit cells along the beam direction in the crystal. Depending on specimen thickness and defocus value, high chemical contrast can be achieved with HRTEM. An example is shown in Fig. 1.2. Theoretical

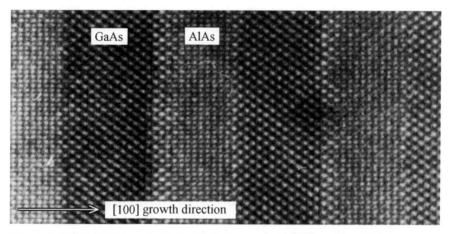

Fig. 1.2. HRTEM image of a GaAs/AlAs heterostructure grown on a (001)-oriented substrate. The image is taken along [$\bar{1}$10]

simulations of the intensities allow a quantitative determination of the sample stoichiometry if the primary beam direction is along $\langle 001 \rangle$ [28]. The results obtained by this method demonstrate that even for optimized growth conditions, significant roughness on an atomic scale is present at the heterointerfaces. The transition layer is typically 2–4 monolayers (ML) thick. For comparison with growth experiments, the [110] and [$\bar{1}$10] directions are more interesting since they correspond to the main axes of anisotropy along the growing (001) surface. The theoretical modeling for these beam directions, however, is much more difficult, since higher-order beams need to be included that depend very sensitively on sample thickness and defocus [29]. This is the reason why quantitative results are difficult to obtain in these cases [30]. Nevertheless, layer thickness fluctuations and terrace width distributions have been determined for the $\langle 110 \rangle$ directions [31,32], confirming the anisotropy of surface islands and steps during GaAs/AlAs epitaxy. Figure 1.2 shows an HRTEM picture of a GaAs–AlAs superlattice (SL) grown by the author. The incident beam direction is along one of the $\langle 110 \rangle$ axes. One can clearly see the differences in abruptness of the AlAs-on-GaAs and the GaAs-on-AlAs interfaces. We shall return to these measurements in Sect. 10.1.1, when we compare them to the results obtained from our RHEED study. Systematic HRTEM studies of the GaAs/AlAs interface reveal that the density of mono-layer steps decreases with growth interruption. The step distributions derived from the SNOM results are thus reproduced. In addition, an intermixing on the atomic scale on the order of a few (001) lattice planes is observed.

True atomic resolution can be achieved by probing the cleaved edge of a heterostructure with a scanning tunneling microscope (STM). For STM, the samples need to be doped to sustain the tunneling current. Both donors and acceptors are then visible in the STM scan as protrusions or depressions of various sizes and obscure the chemical contrast of the heterointerface [33, 34]. To the best of the author's knowledge, only data on GaAs/$Al_x Ga_{1-x}$As ($x \approx 0.3$) alloy heterointerfaces have been published so far [35, 36]. The results indicate a demixing of GaAs and AlAs in the $Al_x Ga_{1-x}$As, forming regions with a typical modulation length scale of 2 nm. This modulation is present at the interface also, causing an interface roughness of the same magnitude. Generally, the scan area of STM is too small to determine the interface struc-ture on a larger scale, restricting measurements of this type to the microscopic domain.

STM scans of the as-grown crystal surface have mainly been published for GaAs [37–39]. The images reveal large atomically flat terraces after growth interruption and annealing, with dispersed anisotropic mounds several layers high. The terrace size of several thousand angstroms corresponds well with the SNOM data. Without annealing, the surface approaches a rougher steady-state structure during growth, with a typical length scale of about 20 nm and a large degree of anisotropy. The root mean squared (RMS) surface height fluctuation of this configuration is about 2.5 Å. STM data of this kind have to be interpreted with some caution as STM generally is an ex-situ technique. The quenching procedure is slow in comparison to the typical timescale of atomic site changes on the surface and therefore the image acquired by the STM may differ from the actual surface structure during growth. Results from pure AlAs surfaces and intermediate stages of interface formation ([40, 41], see Figs. 7.15 and 7.16) indicate a surface morphology similar to that of the pure GaAs surface, with a slightly reduced surface anisotropy. These results are discussed in Sect. 7.2.2.

The structural difference between the $Al_x Ga_{1-x}$As-on-GaAs (normal) and the GaAs-on-$Al_x Ga_{1-x}$As (inverted) interface in MBE-grown heterostruc-tures [42] is of particular importance for device production since the high-frequency performance of devices depends on the electron mobility. In mo-dulation-doped heterostructures using the normal $Al_x Ga_{1-x}$As-on-GaAs in-terface, the electron mobility has been continuously enhanced by refining the growth methods [43, 44]. The best values nowadays peak at more than 10^7 cm^2 V^{-1} s^{-1} below 1 K [45, 46]. Structures of this type are used for fun-damental studies of low-dimensional physics, such as the fractional quantum

Hall effect or ballistic transport. For the inverted (GaAs-on-$Al_xGa_{1-x}As$) interface, experimental electron mobilities are lower by almost an order of magnitude [47,48]. Apart from Si segregation [49], this asymmetry between the normal and inverted interfaces seems to be due mostly to the presence of specific deep-level defects at the inverted interface [50]; see also Sect. 10.4.4. This implies that interface quality cannot be defined by composition variations only. Depending on the type of experiment, the electronic structure of the interface may become important.

Several experiments have been carried out to study the segregation of Ga into AlAs during formation of the normal interface in MBE. X-ray photoemission spectroscopy (XPS) of the surface of continuously grown $Al_xGa_{1-x}As$ layers showed deviations of the surface composition from the stoichiometry of the bulk. The bulk composition was determined by measuring the deposition rates of both GaAs and AlAs separately using RHEED intensity oscillations. The surface composition was found to deviate to the Ga-rich side by up to 25 %, peaking at a substrate temperature of 630 °C ($x = 0.55$) [51]. More detailed ultraviolet photoemission spectroscopy (UPS) experiments performed on nominally abrupt interfaces of AlAs layers grown on GaAs also showed significant intermixing [52]. RHEED-based measurements of transition times between surface reconstructions confirmed the segregation [53]. In these experiments, a few MLs of AlAs were deposited on GaAs followed by sublimation of the remaining surface Ga at elevated temperatures. This sublimation process was monitored by a change in the surface reconstruction. Mass spectroscopy measurements during MBE growth also found significant segregation of Ga into overgrown AlAs layers [54], with a 50 % reduction in signal for every 2 ML of AlAs that is deposited. This exponential decay is independent of growth temperature. Theoretical models for the growth of lateral superlattices indicate the presence of a vertical exchange mechanism between Ga and Al atoms with long-range Ga migration [55].

In summary, the following general features of MBE-grown (001) GaAs/AlAs interfaces can be regarded as well established:

- GaAs/AlAs interfaces are not atomically smooth. Instead, they exhibit a roughness extending several MLs away from the nominal interface position. This roughness is present on different characteristic length scales, depending on the growth conditions.
- AlAs-on-GaAs (normal) and GaAs-on-AlAs (inverted) interfaces are markedly different in both morphology and electronic structure. Ga segregates into the AlAs during formation of the normal interface.

1.3 GaAs/AlAs Surfaces

Although the GaAs (001) surface has been studied for quite some time, there is still considerable uncertainty about the structure of its surface reconstructions. Typical growth conditions on GaAs (001) are ≈ 580 °C surface temper-

8   1. MBE-Grown Semiconductor Interfaces

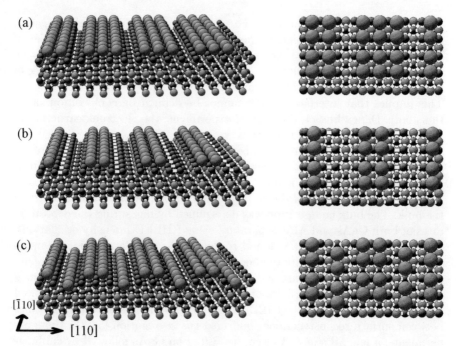

Fig. 1.3. Three structure models for the (001) GaAs (2×4) surface reconstruction. Shown are (**a**) a three-dimer model, (**b**) a two-dimer model with second-layer Ga in the missing dimer rows and (**c**) a two-dimer model with dimerized third-layer As [56]

ature with a V/III flux ratio of about 3. In this range, the surface reconstructs with a (2×4) symmetry. Three different surface reconstructions with this symmetry are observed and are classified according to their RHEED patterns. We follow the nomenclature established by Farrell and Palmstrøm [57]. The $\beta(2\times4)$ reconstruction is characterized by equal intensities in the fractional-order streaks observed in the $[\bar{1}10]$ azimuth. The half-order streak is weaker for both the high-temperature $\alpha(2\times4)$ and the low-temperature $\gamma(2\times4)$ structure. All (2×4) structures are terminated by As dimers, with a missing dimer row along $[\bar{1}10]$. The detailed structure, however, is still a matter of debate. For the $\beta(2\times4)$ structure, the models shown in Fig. 1.3 are the most likely candidates. Figure 1.3a shows a structure with three dimers in the topmost layer; Figs. 1.3b and c show structures with two top-layer dimers and different reconstructions within the missing dimer trench.

The (113)A surface of GaAs/AlAs exhibits special properties that make it an interesting candidate for fundamental studies as well as device fabrication. It is located about halfway between the As-terminated (001) and the Ga-terminated (111)A surface orientation. Although the ideal, unreconstructed GaAs (113)A surface is As-terminated, the first Ga plane is very close to the surface and the top As and Ga planes are almost at the same position.

Both the As and the Ga atoms retain the coordination of the respective extremes, the (001) and the (111)A surfaces. This means that As exposes two dangling bonds and Ga exposes one dangling bond at the surface. Since there are two As dangling bonds for each Ga dangling bond, the unreconstructed surface is polar. The duality of the (113)A surface with respect to Ga and As termination manifests itself in the incorporation of amphoteric dopants as shown in Fig. 1.4. Si becomes a donor on (001) and $(11n)$B (n integer), whereas on (111)A and (112)A, it is incorporated as an acceptor [58].

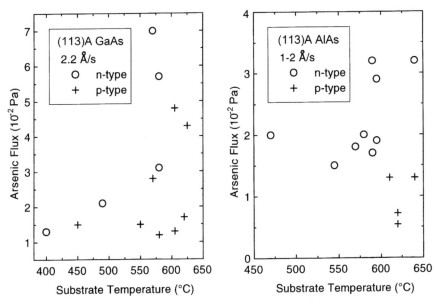

Fig. 1.4. Incorporation behavior of Si as a dopant on GaAs/AlAs (113)A. The doping type can be switched by varying the growth parameters. Data by the author

On (113)A it can be switched between the two types by varying the growth parameters so that either one of the surface properties becomes dominant. This can be achieved by changing the As$_4$ flux during growth [59]. Using this method, p–n junctions can be fabricated using only a Si cell as dopant source [60]. The dopant incorporation behavior is similar for GaAs and AlAs, indicating that it is due to the surface orientation and not a material property.

Two structure models for the GaAs (113)A surface as well as the (hypothetical) unreconstructed surface are shown in Fig. 1.5. The RHEED pattern indicates a surface reconstruction with a period corresponding to eight times the unit of Fig. 1.5c along [$\bar{1}$10]. Because the corresponding RHEED patterns are practically identical (see Fig. 10.8), we can safely assume that the (113)A surface reconstruction structure is the same for both GaAs and AlAs. Once prepared at a substrate temperature of about 600 °C, the reconstruction is stable down to room temperature and up to the onset of sublimation for a wide range of As_4 fluxes. This is a significantly larger window of stability than, for example, for the reconstructions found on the (001) surface.

(a)

(b)

(c)

$[\bar{3}\bar{3}2]$

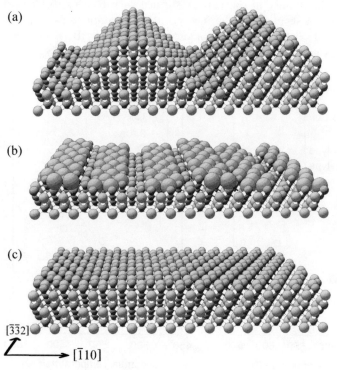

$[\bar{1}10]$

Fig. 1.5. Three-dimensional representations of two surface reconstructions proposed for the GaAs (113)A surface. (**a**) Model proposed by Nötzel et al. [61]. (**b**) Model proposed by Wassermeier et al. [62]. The largest spheres indicate dimerized As atoms. (**c**) Hypothetical unreconstructed surface

On the basis of these observations, a model was proposed by Nötzel et al. [61,63] in which the surface reconstructs into two sets of $(\bar{3}13)$ and $(1\bar{3}3)$ facets running along the $[\bar{3}\bar{3}2]$ direction. The corrugation was explained by a lower surface energy of these facet planes compared to (113)A. A representation of this model is given in Fig. 1.5a. The possibility of lateral quantum confinement in quantum well structures grown on a surface in this orientation sparked widespread interest in the (113) surface [64–70]. Wassermeier et al. [62] proposed a different model of the surface reconstruction on the basis of STM results. This structure model is shown in Figs. 1.5b and 1.6. For this model, the excess electron density and dangling-bond density have been minimized.

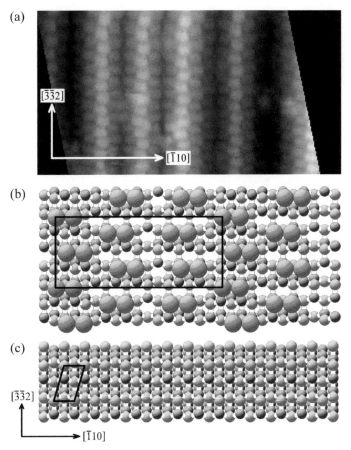

Fig. 1.6. (a) STM scan of the GaAs (113)A surface and (b) top view of the surface reconstruction model proposed by Wassermeier et al. The surface unit cell is indicated by the rectangle. The surface unit cell of the unreconstructed surface is shown in (c)

Figure 1.6 shows a direct comparison of the proposed structure, in top view, and a high-resolution STM scan. The electron-counting model used to determine the structure is based on the assumption that the most stable structure is the one with the lowest excess electron and dangling-bond densities. If a trade-off is possible, the lowest excess electron density is favored. For the Wassermeier model, the excess electron density is zero, while the dangling-bond density assumes a value in between the dangling-bond densities of the commonly accepted models of (001) and (111)A GaAs reconstructions. The model involves only four reconstructed atomic planes, resulting in a depth modulation of 3.4 Å, compared to 10.2 Å for the Nötzel model.

2. Reflection High-Energy Electron Diffraction (RHEED)

Compared to other diffraction methods, the glancing-incidence-angle geometry of RHEED offers many advantages, such as high surface sensitivity and in-situ compatibility with crystal growth. At the same time, however, glancing incidence requires special adaptations of experimental methods and theoretical models to take account of the symmetry breaking introduced by the reflection geometry.

2.1 Geometry and Experimental Conditions

In kinematical scattering theory, the possible reflections are determined by the condition that the wavevectors k_0 and k' of the incident and diffracted beams differ by a reciprocal-lattice vector G:

$$k' - k_0 = G. \qquad (2.1)$$

When considering only elastic scattering events, which means that $|k'| = |k_0|$, this diffraction condition can be cast into the geometrical construction of the Ewald sphere in reciprocal space. In this construction, the tip of k_0 is attached to a reciprocal-lattice point. The sphere around the origin of k_0 with radius $|k_0|$ then defines the Ewald sphere (Fig. 6.2a). Reflections can occur for all k' connecting the origin of the sphere and a reciprocal-lattice point on the sphere. The magnitude of the wavevector for high-energy electrons is given by

$$k_0 = \frac{1}{\hbar}\sqrt{2m_0 E + \frac{E^2}{c^2}}. \qquad (2.2)$$

The relativistic correction amounts to about 3% for $20\,\mathrm{keV}$ electrons. For a qualitative analysis, it is therefore often sufficient to use the nonrelativistic approximation. In RHEED, the Ewald sphere is large; for $20\,\mathrm{keV}$ electrons k_0 is $785\,\mathrm{nm}^{-1}$, which is about 70 times larger than the reciprocal-lattice unit of GaAs. This means that it produces an almost planar cut through the first few Brillouin zones of the reciprocal lattice. This large radius of the Ewald sphere combined with the small scattering angles facilitates the geometrical interpretation of RHEED patterns, since for many purposes angular distortions can be neglected and the usual small-angle approximations for the trigonometric functions are valid.

RHEED is very surface-sensitive in that it samples only very few atomic layers beneath the surface. The surface-normal component k_{0z} of the incident wavevector, which determines the penetration into the material, can be varied over a large range of values by changing the incidence angle θ. For low incidence angles, k_{0z} typically corresponds to energies below $1000\,\text{eV}$, comparable to or even less than the values accessible by low-energy electron diffraction (LEED) [71]. We shall show experimentally that the sampling depth of RHEED in these cases can be very small. This implies that the periodic part of the crystal beneath the surface can usually be neglected. In the construction of the reciprocal lattice we can therefore often approximate the sampled volume by a two-dimensional layer. The reciprocal lattice then degenerates into a set of one-dimensional rods along the z direction perpendicular to the surface. Using this reciprocal lattice, we get the Ewald sphere construction used in RHEED that is shown in Fig. 2.1. Since the reciprocal lattice consists of continuous rods, every rod produces a reflection in the diffraction pattern. The reflections occur on so-called Laue circles of radius L_n centered at H,

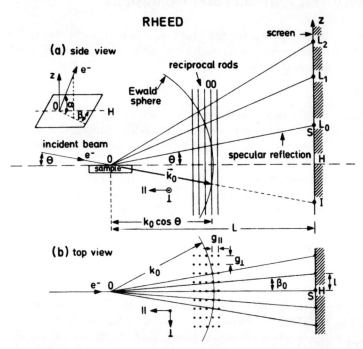

Fig. 2.1a,b. Ewald sphere construction and diffraction geometry of RHEED. Intensity maxima on the screen correspond to projected intersections of the Ewald sphere with the reciprocal lattice. Adapted from Hernández-Calderón and Höchst [72]

the projection of the component parallel to the surface of k_0 onto the screen. The specular reflection or specular spot S is located at the intersection of the zeroth-order Laue circle with the (00) rod. The origin of the reciprocal lattice is projected onto I, also labeled (000), where for some sample geometries the part of the incident beam that misses the sample becomes visible. Owing to the simple geometry of RHEED, the reciprocal-lattice-rod separations $g_\parallel$ and $g_\perp$ parallel and perpendicular to the beam direction can be easily determined:

$$ng_\parallel = k_0 \left[\cos\theta - \frac{1}{\sqrt{(L_n/L)^2 + 1}} \right] , \qquad (2.3)$$

where $ng_\parallel$ is measured from the row containing the (00) rod, and

$$ng_\perp = \frac{k_0}{\sqrt{(L/nl)^2 + 1}} , \qquad (2.4)$$

which for small angles ($nl \ll L$) can be approximated by

$$ng_\perp = \frac{nl}{L} k_0 . \qquad (2.5)$$

Again, $ng_\perp$ is measured from the central rod. The reciprocal-lattice rods are indexed in a way similar to the bulk reflections. The nomenclature is based on the two-dimensional surface lattice, using two indices similarly to the 3D case. The surface lattice vectors generally differ from the bulk lattice vectors so that there is no direct correspondence to the Miller indices of the bulk material. If a surface reconstruction is present that produces a super-period, fractional indices are used to label the superlattice reflections. The center of the reciprocal lattice I, which lies on the zeroth Laue circle by definition, is labeled (000). All angles, however, are measured from the symmetry axis of the zeroth Laue cone O–H.

The reflection geometry introduces a breaking of the crystal symmetry in the z direction. This produces additional effects not present in transmission diffraction. When an electron crosses the crystal–vacuum interface, it gains or loses energy corresponding to the potential difference between the crystal and the vacuum. The literature values for this potential, which corresponds to the zero-order Fourier component of the electronic potential inside the crystal (referred to as the average crystal potential), are around $-14\,\text{eV}$ for GaAs with respect to the vacuum level [73]. The change in energy only affects the surface-normal component of the electron wavevector and therefore is negligible in transmission electron microscopy (TEM) at normal incidence. In RHEED, however, where the surface-normal component of k_0 can be varied continuously from zero, it results, for example, in a distortion of the Kikuchi line pattern close to the shadow edge and a modulation or shift of the intensities in the z direction. These phenomena will be treated in detail in Chaps. 4 and 5, respectively.

Multiple elastic (also called dynamical) scattering generally complicates the analysis since the electron–matter interaction is strong. It occurs when a diffracted beam, in turn, acts as a primary beam for subsequent diffraction. The most important multiple-scattering event is the so-called surface resonance effect [74–77]. This is characterized by a diffracted beam traveling parallel to the sample surface. This condition is met when the Laue circle tangentially touches a reciprocal-lattice rod. The k-vector of this wave along the surface defines the same Ewald sphere as k_0 and and can therefore transfer intensity to any of the reflections on the Laue circles.

Inelastic scattering can take place via several different mechanisms [78, 79]. *Plasmon excitation* typically involves energy losses of 10 to 30 eV with an angular spread of the scattered electrons of $\theta_E = \Delta E/2E_0$, where ΔE is the electron energy loss and E_0 is the incident electron energy. In this process, the incident electron excites charge density waves in the ensemble of loosely bound electrons within the solid or at the crystal surface. Depending on the incidence angle, bulk and surface plasmons, which can be distinguished by their different excitation energies, are created in variable amounts. Plasmon inelastic scattering is strongly favored for small momentum transfers; typical values of the momentum transfer are on the order of 0.1 %. Therefore, changes in the RHEED pattern due to plasmon scattering can only be resolved with very-high-resolution instruments [80]. For larger momentum transfers, the plasmon dispersion merges with the electron–hole excitation continuum. *Band-to-band transitions* are also excited at smaller energies. For semiconductors, the transitions corresponding to the band gap of the material can be strong, splitting into several peaks in large-band-gap materials and insulators. *Atomic core-shell excitations* involve an energy loss of a few hundreds to thousands of eV and an angular spread similar to plasmon loss events ($\approx \Delta E/2E_0$). At high energies, atomic core-shell excitations become less probable and, owing to their large range of energy and momentum transfer values, contribute to a featureless background. Theoretical arguments indicate that plasmon scattering is the dominant inelastic scattering event in RHEED, with a typical electron mean free path of several hundred angstroms at 10 keV in metals [81]. This means that it is of the same order of magnitude as the elastic scattering. Energy-filtered RHEED experiments confirm these arguments; see Chap. 8. The effect of plasmon scattering is to broaden the reflection profiles at low intensities. In high-resolution peak profile analysis, this broadening has to be taken into account [80]. *Phonon scattering* causes a large momentum transfer, but an energy loss of typically only a few meV. Although energy filtering with meV resolution does not seem feasible in combination with MBE, the energy loss of phonon-scattered electrons destroys coherence. Diffraction features due to phonon inelastic scattering should therefore be independent of the elastic scattering pattern.

2.2 Instrumentation
and Miscellaneous RHEED Techniques

A typical RHEED measurement system consists of an electron gun, a phosphor screen and image-processing hardware and software. The electron gun is designed with a typical focal length of about 0.5 m, combined with a very low divergence of the beam. This ensures a small spread of diffraction conditions for the electrons and the sampling of a small and therefore relatively homogeneous area of the sample. In the ideal limit, the beam consists of electrons that propagate in the same direction with the same energy and hit the sample at the same location. Typical acceleration voltages range from 10 keV to 30 keV. This high energy is necessary to image a sufficient area of reciprocal space into the relatively small solid angle of the phosphor screen. It has the additional advantage of decreasing the influence of stray fields, which are abundant because of the electric heaters, electric motors and magnetic vacuum feedthroughs. The electron beam touches the sample at an angle that is typically in the range of 0.5–2.5°. The diffracted intensity pattern is then converted to visible light by a phosphor screen. The chamber side of the screen is coated with a metal layer to block ambient light and low-energy electrons. An electron hitting the screen at 20 keV typically produces 1000 photons in the green part of the optical spectrum. A good way to analyze the entire area of a RHEED pattern with the necessary time resolution is through detection with a charge-coupled device (CCD) camera outside the UHV. This allows sampling frequencies of typically 25 Hz. To record and process several hundred intensity values with this frequency, powerful image-processing systems have been developed [82]. Several of these PC-based hardware and software packages are commercially available.

An alternative approach to recording the complete diffraction pattern is a scanning technique using a single-channel detector. To obtain reflection profiles, either the detector is moved across the phosphor screen [83] or the RHEED pattern is deflected by means of magnetic [84] or electric [85, 86] fields in the vacuum. This approach offers the advantage of using high-sensitivity detectors with large dynamical range and is therefore ideal for the measurement of static beam profiles. If the detector is mounted inside the vacuum chamber, the spatial resolution can be made almost arbitrarily small while at the same time allowing the addition of an energy filter [85,86]. The main disadvantage of scanning techniques is their long acquisition times. For the measurement of RHEED dynamics, CCD detection using a phosphor screen is still, therefore, the method of choice. Instruments combining the speed of CCD detectors with an energy-filtering capability will be discussed in Chap. 8.

To study special problems or to extend the capabilities of RHEED, dedicated setups have been constructed. These experimental setups are usually not compatible with standard MBE growth chambers and are therefore only briefly reported here, without any claim of completeness. RHEED can also be used in reactors with high process pressures up to 10^{-3} mbar by differentially pumping the electron gun and operating it at a higher vacuum level. One nice example is the study of catalytic CO oxidation at Pt (110) surfaces, featuring oscillations driven by the surface reaction within a narrow O and CO pressure range [87]. Two RHEED guns have been mounted on the same MBE chamber to record diffraction patterns and RHEED dynamics in two azimuths simultaneously [88,89]. This allows the study of growth modes and phase transitions on anisotropic surfaces with high temporal resolution. Microprobe RHEED [90–92] is a combination of scanning electron microscopy (SEM) and RHEED, where an SEM gun is used as the electron source, allowing the simultaneous real-space imaging of the surface with a spatial resolution of a few microns. An ingenious way of directly recording the reciprocal surface structure mesh using the geometry of Fig. 2.1 was developed by Ino [93], using a spherical phosphor screen with its radius centered at the origin O on the sample. This screen is then imaged along the z direction of Fig. 2.1, directly showing a compressed image of the 2D reciprocal lattice similar to LEED. Another way of directly recording the surface reciprocal lattice, by rotating the sample, is presented in Chap. 6.

Finally, there are several miscellaneous RHEED applications that are not treated in detail in this book, but address phenomena frequently encountered in MBE growth. They are therefore of interest to crystal growers using RHEED. The following is a brief discussion of a few references, again without a claim of completeness.

While RHEED oscillations are usually used to determine the flux of the rate-limiting species in MBE, the same technique can be modified to measure fluxes of species that are supplied at an overstoichiometric rate during growth. This is usually done by depositing only the rate-limiting species for a certain time to form an accumulation layer, which is then consumed when the supply is switched to the flux of the second source [94]. The background level of the measured species can be determined at the same time, and modified versions of the measurement have been developed [95] that rely on the disappearance of the oscillations when the flux ratio is inverted. This method, when applied to As, is commonly referred to as As-induced oscillations.

Transitions between single- and double-period RHEED oscillations are observed in several materials when the growth parameters are varied [96, 97]. These transitions, however, can be due to changes in the surface reconstruction and the accompanying change in the shape of the oscillations (see Sects. 9.2.2 and 10.2) instead of an actual transition between single- and double-layer growth. To rule out diffraction-related effects, the period-doubling transition has to be independent of diffraction conditions, which means it has to be independent of the choice of incidence angle and azimuth [98].

Stranski–Krastanov or Volmer–Weber growth of lattice-mismatched materials leads to three-dimensional growth islands, which are used to form quantum dots. The formation [99] and crystallographic orientation [100] of these mounds can be controlled with RHEED. The growth of the islands results in a relaxation of the lattice constant, which can be followed by measuring the spacing of the corresponding transmission diffraction spots [101].

Polycrystalline samples can also be investigated by RHEED [102,103]. By fitting the shape of the diffraction pattern with a kinematical model, the texture of the film, the average angle of the texture axis with respect to the surface normal and the angular width of the distribution of this angle about its average value can be determined.

2.3 Theoretical Models

RHEED theory starts from the following Schrödinger equation, which is valid at beam energies where relativistic corrections are small:

$$\left[\nabla^2 + U(\boldsymbol{r}) + \boldsymbol{k}_0^2 \right] \Psi(\boldsymbol{r}) = 0 \,. \tag{2.6}$$

Here, $\boldsymbol{k}_0$ is the (relativistic) incident wavevector in the vacuum and $U(\boldsymbol{r})$ is a scaled lattice potential given by

$$U(\boldsymbol{r}) = \frac{2me}{\hbar^2} V(\boldsymbol{r}) \,, \tag{2.7}$$

where m is the (relativistic) electron mass, $-e$ is the electron charge and $-eV(\boldsymbol{r})$ is the potential energy of the electron inside the crystal. We consider only elastic scattering, $|\boldsymbol{k}_0| = |\boldsymbol{k}'|$, with $\boldsymbol{k}'$ being the wavevector of the scattered wave. Equation (2.6) can then be rewritten in integral form as

$$\Psi(\boldsymbol{r}) = \exp(\mathrm{i} \boldsymbol{k}_0 \cdot r) + \int G(\boldsymbol{r}, \boldsymbol{r}') U(\boldsymbol{r}') \Psi(\boldsymbol{r}') \mathrm{d} \boldsymbol{r}' \,, \tag{2.8}$$

where the first term represents the plane incident wave and the integral determines the scattered wave. One form of the Green function G is

$$G(\boldsymbol{r}, \boldsymbol{r}') = \frac{\exp(\mathrm{i} k_0 |\boldsymbol{r} - \boldsymbol{r}'|)}{4\pi |\boldsymbol{r} - \boldsymbol{r}'|} \,, \tag{2.9}$$

giving the amplitude at a point $\boldsymbol{r}$ due to a point of unit scattering strength at $\boldsymbol{r}'$. In order to obtain a rough estimate of the scattering cross-section, we consider the potential of a single atom with Thomas–Fermi screening [104],

$$V(r) = -\frac{Ze^2}{4\pi \varepsilon_0 r} \exp\left(-\frac{r}{a}\right) \,, \tag{2.10}$$

where the screening length a is given by

$$a = \frac{4\pi \varepsilon_0 \hbar^2}{me^2 \sqrt[3]{Z}} \,, \tag{2.11}$$

with Z being the atomic number of the atom and ε_0 the permittivity of free space. In the Born approximation the total cross-section then is [104]

$$\sigma = \frac{16\pi m^2 e^4}{\hbar^4 (4\pi\varepsilon_0)^2} \frac{Z^2 a^4}{4k^2 a^2 + 1}. \tag{2.12}$$

For the As atom ($Z = 33$) as a typical example, σ is about $2.5\,\text{Å}^2$ at $20\,\text{keV}$; if a disk of area σ were placed at each atomic position, roughly $15\,\%$ of the area of a (001) plane in GaAs would be covered. This means that the kinematical treatment, where only single scattering events are considered, cannot be expected to be very accurate in the RHEED case. Owing to the small angle of incidence, the interaction volume is strongly elongated in the beam direction and therefore multiple scattering is very likely. This is the reason why theoretical treatments of RHEED generally include multiple scattering. On the other hand, these more comprehensive calculations are expensive in computer time and not very intuitive, in that the occurrence of any particular intensity variation in the calculated reflection pattern cannot be easily linked with a feature of the surface structure. Even sophisticated calculations are hampered by convergence problems and often do not agree well with the experimental data. There is an ongoing debate as to whether this type of RHEED theory can be regarded as a reliable technique or should be considered to be in its infancy [105–107]. Kinematical models are therefore still used for the interpretation of RHEED data. Despite their obvious shortcomings, they offer two big advantages, namely the simplicity of the calculation and the intuitive and instructive dependence of the results on the input structure.

2.3.1 Kinematical Scattering

Starting from (2.8), we can approximate the wave function inside the crystal by the incident plane wave, which gives an equation identical to that of first-order perturbation theory. The scattered wave then becomes

$$\Psi(\boldsymbol{r}) = \frac{1}{4\pi} \int \frac{\exp(\mathrm{i}k|\boldsymbol{r} - \boldsymbol{r}'|)}{|\boldsymbol{r} - \boldsymbol{r}'|} U(\boldsymbol{r}') \exp(\mathrm{i}\boldsymbol{k}_0 \cdot \boldsymbol{r}') \mathrm{d}\boldsymbol{r}', \tag{2.13}$$

and the amplitude for scattering into the final direction $\boldsymbol{k}'$ is given by

$$F(\boldsymbol{k}_0 - \boldsymbol{k}') = \frac{1}{4\pi} \int \exp[\mathrm{i}(\boldsymbol{k}_0 - \boldsymbol{k}') \cdot \boldsymbol{r}'] U(\boldsymbol{r}') \mathrm{d}\boldsymbol{r}'. \tag{2.14}$$

We express the scattering potential $V(\boldsymbol{r})$ as the sum of the contributions of individual atoms

$$V(\boldsymbol{r}) = \sum_i \phi_i(\boldsymbol{r} - \boldsymbol{r}_i) \tag{2.15}$$

and insert the Fourier components of this potential

$$u_i(\boldsymbol{k}_0 - \boldsymbol{k}') = \int \phi_i(\boldsymbol{r}) \exp[\mathrm{i}(\boldsymbol{k}_0 - \boldsymbol{k}') \cdot \boldsymbol{r}] \mathrm{d}\boldsymbol{r} \tag{2.16}$$

into (2.14) to obtain

$$F(\mathbf{k}_0 - \mathbf{k}') = \frac{me}{2\pi\hbar^2} \sum_i u_i(\mathbf{k}_0 - \mathbf{k}') \exp[\mathrm{i}(\mathbf{k}_0 - \mathbf{k}') \cdot \mathbf{r}_i] \,. \tag{2.17}$$

The potential components u_i are generally given in modified form as a function of the parameter s, defined by

$$s = \frac{\sin\theta}{\lambda} \,, \tag{2.18}$$

where λ is the incident electron wavelength and θ is the angle between $\mathbf{k}'$ and $\mathbf{k}_0$. These components can be found in tabulated form for atoms, calculated by the Hartree–Fock method [108]. A graphical representation of these so-called Doyle–Turner potentials for Al, Ga and As is given in Fig. 2.2. Whereas the scattering factors for Ga and As are very similar, the Al scattering factors are markedly smaller. GaAs therefore is very close to the diamond structure, and the zinc blende reflections are quasi-forbidden. The distinct difference between Al and Ga is the reason for the relatively good material contrast in TEM, as, for example, in Fig. 1.2. It will also become important when we discuss the sampling depth of RHEED in Sect. 7.2.1. Note that the expansion in Gaussians used in the calculations described later deviates significantly from the tabulated values for large s, which means that intensities far from the origin of the reciprocal lattice are too strongly damped in the simulations. We will return to this problem in Sect. 4.3.

Taking further advantage of the crystal periodicity, we can introduce the reciprocal-lattice vectors $\mathbf{G}$ to obtain the scattering amplitude

$$f(\mathbf{G}, s) = N\frac{me}{2\pi\hbar^2} \sum_i u_i(s) \exp[-B_i(T)s^2] \exp(\mathrm{i}\mathbf{G} \cdot \mathbf{r}_i) \,, \tag{2.19}$$

where N is the number of unit cells in the interaction volume, the reciprocal-lattice points $\mathbf{G}$ define the possible reflections, and i runs over the basis atoms in one unit cell. The intensity of the reflection labeled by the integer indices of $\mathbf{G}$ is then determined by $|f(\mathbf{G})|^2$. The reduced intensity of the elastic reflections due to thermal motion of the atoms can be included to a good approximation by means of the Debye–Waller factors. These prefactors to the exponential in the equation are of the form $\exp[-B_i(T)s^2]$ and therefore of the same shape as the Doyle–Turner potential. If relative intensities are considered, these factors can generally be neglected. The $B_i(T)$ are also available in tabulated form for a variety of materials [109]. Equation (2.19) describes bulk materials. Modifications for the 2D case will be discussed in Chap. 4.

The kinematical treatment of RHEED is valid for the determination of the surface unit mesh as in Fig. 2.1, which involves the determination of reflection positions parallel to the surface. This includes reflection profile analysis parallel to the surface, if plasmon losses are taken into account. These usually only affect the beam profile at small intensities [80] and can be neglected in most cases. For the interpretation of intensities and intensity modulations in the z direction perpendicular to the substrate surface, the application

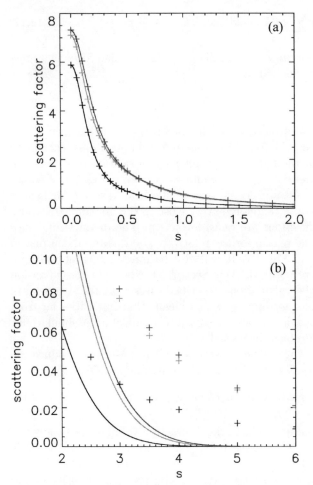

Fig. 2.2. Calculated electron scattering factors for Al (*black*), Ga (*light gray*) and As (*dark gray*) using the Hartree–Fock method [108]. The tabulated values are given as *crosses*, whereas the *lines* represent the expansion in Gaussians used in the calculations. The fit distinctly deviates from the tabulated values for large s

of a purely kinematical treatment does not generally lead to good agreement and a careful estimate of its reliability must be performed. Kinematical treatments have been mainly used to study the effects of surface disorder and absorption on the RHEED pattern [110–112] and to interpret diffraction from stepped systems [113, 114]. Comparisons of the kinematical treatment with dynamical calculations show that a kinematical analysis of superlattice intensities is possible if no strong fundamental reflections are excited [115, 116]. Such kinematical analyses of RHEED intensities have been performed for Si and GaN surface reconstructions [117–119]. In Sect. 4.3 we shall show that modifications of the kinematical treatment to take account of the average crystal potential can significantly improve the agreement between theory and experiment.

The simplicity and power of the kinematical treatment rely on the Fourier-transform correspondence between real and reciprocal space. Since the scattering amplitude is proportional to the Fourier transform of the real-space potential, a reconstruction of the real-space lattice from the intensity distribution would be possible if the scattering amplitude could be measured directly. We can, however, measure intensities only. Owing to the loss of the phase information in the intensity (called the phase problem), the inverse Fourier transform of the intensity distribution does not reproduce the potential distribution of the surface. Instead, according to the convolution theorem of Fourier transformation, we obtain its autocorrelation function:

$$2\pi |A(\boldsymbol{k})|^2 = 2\pi A(\boldsymbol{k})A^*(\boldsymbol{k}) = F\left[\int_{-\infty}^{\infty} a(\boldsymbol{r}')a(\boldsymbol{r} - \boldsymbol{r}')\mathrm{d}\boldsymbol{r}'\right] = F[a*a] \quad (2.20)$$

In this expression, $A(\boldsymbol{k})$ is the Fourier transform of the real function $a(\boldsymbol{r})$, the symbol $*$ denotes the convolution operation, $\boldsymbol{k}$ is the momentum transfer or reciprocal-space coordinate, F is the Fourier transform operator and the right-hand side defines the autocorrelation function of a.

For a perfect surface, the resolution of RHEED is limited only by the quality of the electron beam. This instrumental limitation can be expressed mathematically by the instrument response function $T(\boldsymbol{k})$, which represents the intensity function the instrument would record for diffraction from a perfect and infinitely large surface [120]. With $I(\boldsymbol{k})$ denoting the diffraction pattern of the surface using an ideal source, we can write down the observed intensity distribution $J(\boldsymbol{k})$ as

$$J(\boldsymbol{k}) = I(\boldsymbol{k}) * T(\boldsymbol{k}) = F[\phi(\boldsymbol{r})] * F[t(\boldsymbol{r})] = F[\phi(\boldsymbol{r})t(\boldsymbol{r})], \quad (2.21)$$

where $\phi(\boldsymbol{r})$ is the autocorrelation function and $t(\boldsymbol{r})$ is the inverse Fourier transform of the instrument response function, called the transfer function. This transfer function is useful for assessing the surface properties that can be represented in a RHEED pattern. Since $T(\boldsymbol{k})$ can be approximated by a Gaussian function, so can $t(\boldsymbol{r})$. The transfer function represents a window of width t_{w}, the transfer width [121]. The information contained in RHEED is therefore random and average in character, with a limited detectable correlation length determined by the instrument response. The transfer width can be determined in the limit of a perfect crystal from

$$t_{\mathrm{w}} = \frac{a_\perp g_\perp}{h_{\mathrm{t}}}, \quad (2.22)$$

where h_{t} is the full width at half maximum (FWHM) of the reflection, and $a_\perp$ and $g_\perp$ represent the real and reciprocal surface lattice constants parallel to the screen. For typical surfaces investigated in this type of work, disorder is significant. Therefore, instrumental limitations are smaller than surface-structure-induced broadening, as can be verified by inspection of Fig. 2.3.

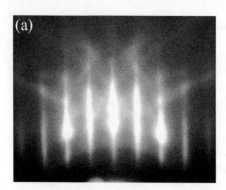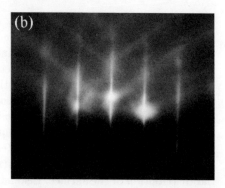

Fig. 2.3. RHEED patterns showing reflection widths for (**a**) a typical $c(4\times4)$ AlAs surface and (**b**) a (110) surface cleaved in UHV. The cleaved surface is well ordered and can therefore be used to determine a lower bound for t_w. Both pictures were taken with the same instrument at 20 kV and are reproduced at the same scale

The transfer width is strongly anisotropic with respect to the beam direction. By varying the incidence angle and interpolating the obtained beam profiles perpendicular to the beam and surface, one can obtain an intensity profile of a reciprocal-lattice rod along the beam direction. Comparison of the FWHM of the rod along and perpendicular to the beam direction yields lower bounds for t_w of typically 10 000 Å and 300 Å in the two directions [122]. Good instruments, even at 6.1 keV, reach 20 000 Å and 2000 Å, respectively [80,85].

Since the diffracted intensity for a perfect instrument is the Fourier transform of the autocorrelation function, the problem of the kinematical scattering theory for disordered surfaces can be reduced to the determination of the autocorrelation function. Quite sophisticated treatments can be found in the literature [123,124]. For our purposes, however, the qualitative behavior and the limiting cases are sufficient.

The reciprocal lattice parallel to a surface with lateral roughness on all scales consists of a constant diffuse background, which represents a continuous spectrum of spatial frequencies. In addition, a sharp δ-like central peak is present, which is due to the fact that the autocorrelation function always has a maximum at zero. This spike corresponds to the (000) or specular spot in RHEED. For a perfectly periodic surface, on the other hand, the autocorrelation function is also perfectly periodic and the corresponding reciprocal lattice consists of a regularly spaced array of sharp peaks. Any arrangement intermediate between these two extremes generally contains both types of contribution, i.e. more or less sharply defined peaks due to periodic contributions and a diffuse background intensity that reflects disorder such as the surface step distribution. A thorough discussion of autocorrelation functions for various structures and the corresponding reciprocal-space representations in the kinematical approximation can be found in the book by J. M. Cowley [125].

2.3.2 Dynamical Scattering

At the moment, there are mainly three methods commonly adopted to solve the problem of computing the full wave function inside the crystal. For a list of references, see [126]. Using the standard approach of solid-state physics, all of these methods use Fourier transformation to take advantage of the periodicity of the crystal potential in some way.

The first approach, dating back to the 1930s, is called Bethe's or the Bloch wave method [127, 128]. It involves the full three-dimensional Fourier transform of both the crystal potential and the wave function. This method is commonly used in TEM, where only a limited number of Fourier coefficients are important. In the RHEED case, it is less practical since the strongly varying potential at the crystal surface requires the inclusion of many Fourier coefficients [129].

One can therefore carry out the Fourier transform in the two dimensions parallel to the crystal surface, in which perfect periodicity is assumed, and retain the real-space coordinate in the perpendicular z direction. This is called the parallel-slicing method. The Fourier transform of the potential in this case is given by [130]

$$V(\boldsymbol{r}) = V(\boldsymbol{r}_t, z) = \sum_m V_m(z) \exp(\mathrm{i}\boldsymbol{G}_m \cdot \boldsymbol{r}_t), \tag{2.23}$$

where $\boldsymbol{r}_t$ is a two-dimensional vector in the surface plane and the $\boldsymbol{G}_m$ represent the complete, two-dimensional set of reciprocal-lattice vectors for the periodic surface. The total wave function is similarly transformed, and we obtain

$$\Psi(\boldsymbol{r}) = \sum_m c_m(z) \exp[\mathrm{i}(\boldsymbol{k}_{0t} + \boldsymbol{G}_m) \cdot \boldsymbol{r}_t], \tag{2.24}$$

with $\boldsymbol{k}_{0t}$ denoting the surface-parallel component of the incident wavevector. By substituting (2.23) and (2.24) into the Schrödinger equation (2.6), we obtain

$$\frac{\mathrm{d}^2}{\mathrm{d}z^2} c_m(z) + \Gamma_m^2 c_m(z) + \sum_n V_{m-n}(z) c_n(z) = 0, \tag{2.25}$$

where Γ_m is the surface normal of the diffracted wavevector for the mth rod. Now the problem is reduced to the integration of this set of coupled equations in the single variable z. To do this, the crystal is divided into thin slices perpendicular to z, with the application of the appropriate boundary conditions. The calculation generally diverges and special measures have to be taken to ensure convergence. This method is an adaptation of LEED calculation methods to the RHEED case.

The third method is called the multislice formalism [131] or perpendicular-slicing method [132]. In this type of calculation [133–135], the unit cell of the interaction volume is sliced into thin layers perpendicular to the sample surface and the beam direction. This is a real-space method in that only the periodicity of the problem along the incident beam direction can be employed for simplification. This periodicity condition can be written down as

$$\Psi(r) = \Psi(r_{\mathrm{p}}, r_{\|}) = \Psi'(r_{\mathrm{p}}) \exp(\mathrm{i}\alpha r_{\|}'), \tag{2.26}$$

where r_{p} is a vector in the plane perpendicular to the incident beam containing the surface normal, and $r_{\|}$ denotes the coordinate parallel to the incident beam direction along the surface. This is then used, together with the boundary condition of matching at the interfaces between slices, for an iterative algorithm using the Green function formulation of the Schrödinger equation (2.8)

$$\Psi_n(r) = \exp(\mathrm{i}k_0 \cdot r) + \int G(r, r')U(r')\Psi_{n-1}(r')\mathrm{d}r', \tag{2.27}$$

where $\Psi_n(r)$ is the nth-order approximation to $\Psi(r)$ and the wavefield is assumed to converge in successive iterations. Generally, it does not, and corrections must be introduced to compensate for the truncation of the wave function at the edge of each slice, which has been identified as the origin of the divergence [136]. The advantage of this method is its capability to include arbitrary deviations of the potential at the crystal surface along the beam direction.

All three methods of dynamical RHEED theory share the difficulties associated with discretization and truncation often inherent in numerical treatments of this complexity. In general, different methods applied to the same problem do not yield the same result [136], although efforts are under way to identify possible sources of the discrepancies [132]. As an example illustrating the general accuracy of the different methods, three calculations for the GaAs $\beta(2\times4)$ surface reconstruction find good agreement with the three-dimer model [105, 137, 138], although predominantly the two-dimer structure has been found in atomic resolution STM studies [56, 139, 140]. Ab initio calculations also confirm the two-dimer model [141, 142]. Dynamical RHEED theory therefore does not seem to be a very reliable tool yet. Compared to the kinematical approximation, the dynamical models generally agree better with the experiments, but not dramatically so. One of the reasons for the rather poor ability of both kinematical and dynamical models to describe scattering from real surfaces is certainly the presence of disorder at any crystal surface, which may drastically modify the RHEED pattern. Disorder is difficult to treat in a realistic way using the present schemes that rely on the strict periodicity of the surface in the directions in which the Fourier transformation is carried out. The treatment of realistic surface disorder within the dynamical scattering framework is a very active area of current research, and promising approaches to solve the problem are being developed as this book is being printed [143–147].

3. RHEED Oscillations

Since they were first reported in 1980 [148–150], RHEED oscillations have been intensively studied since they convey in-situ and real-time information on the dynamics of MBE growth [84]. This allows their use not only as an analytical tool, but also for real-time feedback control in, for example, phase-locked epitaxy [151]. Usually, the specular spot intensity is plotted as a function of time.RHEED oscillations are used to determine growth rates, layer thicknesses and alloy compositions, since it is well established through comparison with other methods that theperiod of an oscillation corresponds to the growth of one ML of GaAs or AlAs [152].

3.1 Current Experimental Status

RHEED oscillations are generally observed under growth conditions that lead to layer-by-layer growth. In this growth mode, also called Frank–van der Merwe growth, one layer is essentially completed before material is added to the following layer. This periodic variation of the surface morphology is generally accepted as the reason for RHEED oscillations. When analyzing experiments, it is useful to distinguish between growth-induced and diffraction-induced effects. This can usually be done either by changing the diffraction conditions in repeated measurements with the same growth parameters or by changing the growth conditions for constant diffraction conditions.

From a RHEED intensity oscillation measurement like the one shown in Fig. 3.1, several parameters thatcharacterize the curve can be extracted and analyzed. Among them are the period, amplitude, phase and damping of the oscillations, the behavior at the initiation of growth, the recovery after growth and the frequency distribution in the Fourier spectrum of the oscillations. In favorable cases, a study of these parameters as a function of either growth or diffraction parameters allows the analysis of growth mechanisms as well as the nature of the diffraction. We briefly review some important results.

During layer-by-layer growth in MBE, the surface periodically changes its morphology because of the nucleation and coalescence of islands in the growing layers. The damping of the oscillations and the recovery after growth both happen on a timescale considerably longer than the oscillation period. We can interpret this as an increase of long-range (long-spatial-wavelength)

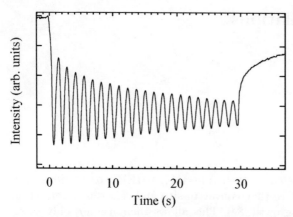

Fig. 3.1. Typical RHEED intensity oscillations for GaAs growth on GaAs (001). The observation is in the [$\bar{1}$10] azimuth at the specular spot position for an incidence angle of 0.9° with an electron energy of 20 keV. The surface reconstruction is $\beta(2\times4)$

roughness that approaches a steady state in the limit of totally damped oscillations. After growth interruption, the surface again smooths towards the thermodynamic equilibrium state. This interpretation is confirmed by STM studies of GaAs surfaces quenched at different stages of growth [38, 153]. The slowly varying roughness, similarly to the discussion in Sect. 1.2, can be regarded as a low-frequency contribution to the surface roughness spectrum, whereas the roughness variation responsible for the RHEED oscillations constitutes a high-frequency component that depends on the position within an ML growth cycle.

As soon as the surface undulation due to the long-wavelength roughness creates average terrace sizes comparable to the mean nucleation distance of layer-by-layer growth, the oscillations are damped out. Every deposited atom reaches a step edge before forming an island nucleus, and growth proceeds, at least locally, by lateral translation of the terrace distribution. Now the autocorrelation function and its Fourier transform, the RHEED intensity, are stationary with respect to the high-frequency component and no oscillations are detected.

This notion can be verified by measuring RHEED oscillations on vicinal surfaces. When the growth temperature and with it the surface mobility of the adatoms is increased, the amplitude of the RHEED oscillations decreases continuously until the oscillations disappear. This transition from layer-by-layer to step-flow growth has been used to obtain estimates of the adatom surface mobilities. The basic idea of these estimates is to compare the RHEED oscillations to surface morphology simulations obtained by Monte Carlo methods [154, 155].

The fast and slow components in the recovery of the intensity after growth interruption [156] can therefore be related to the amplitude of the oscillations at growth termination and the damping envelope of the oscillations, respectively. The fast component in the recovery should depend on the position within the ML growth cycle, whereas the slow component should not.

Damping of the oscillations is also influenced by geometrical factors. Since the flux density of a Knudsen cell decreases as $1/r^2$ and the cells are typically installed at an angle of at least 10–20° to the sample normal, the rate of arrival of atoms is nonuniform across the sample. Owing to the low incidence angle and non-negligible beam diameter, the electrons average over typically 10 mm of sample surface along the beam direction and therefore produce a signal that corresponds to the inverse Fourier transform of a range of frequencies. Such a signal has the form

$$I(t) = \frac{1}{t}[\sin(\omega + \Delta\omega)t + \sin\omega t]\,, \tag{3.1}$$

a damped wave with beating corresponding to the frequency difference $\Delta\omega/2\pi$. Oscillations of this type are shown in Fig. 3.2 as numerical inverse Fourier transforms for $\Delta\omega/\omega = 8\,\%$. This is a common value for standard

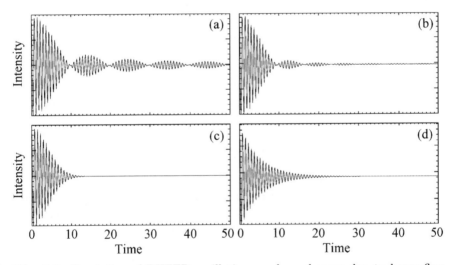

Fig. 3.2. Simulation of RHEED oscillation envelope changes due to beam flux nonuniformity. The curves were obtained by inverse Fourier transformation of (**a**) rectangular, (**b**) parabolic, (**c**) Gaussian and (**d**) Lorentzian frequency distributions

MBE chamber geometries. If the frequency distribution is not a rectangle, but a parabola or a Gaussian or Lorentzian curve of the same height and area, we obtain the results in Figs. 3.2b–d. Owing to the high- and low-frequency tails of the Gaussian and Lorentzian distributions, the beating is suppressed and the damping is much stronger. In a typical experiment, the frequency distribution will most likely assume a shape intermediate between the rectangular and bell-shaped forms, closer to the parabolic distribution.

Whenever the frequency distribution has a sharp cutoff, as for the parabolic distribution, beats are present. Therefore, the shape of the damping envelope is difficult to interpret in a unique manner since it depends strongly on the beam flux nonuniformity. One possible approach to correlating the damping envelope with the long-wavelength roughening during growth is to compare the damping with the shape of the recovery curve after growth. If the fast component of the recovery is much larger than the oscillation amplitude at growth interruption, the damping is due to nonuniformity effects; if it is of the same magnitude, the damping is due to increasing roughness. A typical experiment will include contributions from both effects. Spatial flux nonuniformity effects can be suppressed by using small sample sizes, down to 1 or 2 mm.

Beats similar to those in Fig. 3.2 were observed experimentally [157], and their geometrical origin was demonstrated by adding a flux from a second cell installed at the opposite angle, after which the beating disappeared and the damping was reduced significantly [158]. RHEED oscillations can also be observed in the specular spot with rotating samples. If the rotation frequency is close to the RHEED oscillation frequency, Fourier techniques can be employed to separate the two components. Such experiments confirm the connection between damping and flux nonuniformity [159]. For fast rotation, the damping is significantly smaller than for a static wafer position [160,161]. RHEED with azimuthal rotation is treated in detail in Chap. 6.

Beating, however, is not necessarily due to macroscopic effects. Figure 3.3 shows RHEED oscillations of AlAs growth on AlAs, recorded simultaneously on the (00) and (01) streaks at the intersection with the Laue circle. Whereas strong beating with complete extinction at the nodes is observed on the central streak, the (01) oscillations show a smoothly decaying envelope. At the same time, recovery is fast, indicating only small long-scale roughening during growth. If the beating were due to flux variations, it would be present at all oscillating regions of the RHEED pattern since the diffraction conditions are practically identical at all points of the elongated beam–sample interaction volume. The beats must therefore be of microscopic origin. This means that two different microscopic processes produce surface morphology variations of different frequency. These two microscopic processes contribute differently to the {00} and {01} beams, producing the observed pattern.

We speculate that the anisotropy of the (001) surface might lead to a difference of growth processes in perpendicular directions along the surface. Rather abstract theoretical studies predict temporally periodic phases for two dimensions and anisotropic rules [162]. The microscopic beating is quite an elusive phenomenon. So far, we have been unable to find reproducible conditions for the appearance of these beats. They seem to depend sensitively on either the composition of the residual gas or the substrate strain, or both. The sample of Fig. 3.3 featured a thick buffer of various AlAs–GaAs superlattices. Beats have also been reported for GaAs growth on surfaces modified

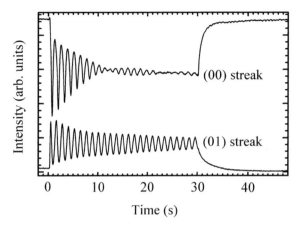

Fig. 3.3. Beating of microscopic origin. Whereas beats with complete extinction at the nodes are observed on the central streak, practically no beating is seen on the first-order streak. This behavior cannot be explained by macroscopic flux variations

by predeposited Sn, where the Sn was incorporated between experiments by overgrowth at low temperatures [163], and an explanation has been presented which relies on the formation of mesoscopic surface undulations that lead to locally different growth rates [164]. A similar case is discussed in Sect. 10.3 for carbon doping.

The intrinsic damping of RHEED oscillations can be quite small. The flux-nonuniformity damping can be minimized by using small samples and fluxes from cells at opposite positions. If the geometrical damping is suppressed, oscillations can be sustained for extended periods of time. The oscillations of Fig. 3.4 were measured on a sample of size 2×2 mm. In this favorable case, the growth of a $0.3\,\mu$m thick layer could be monitored using RHEED oscillations. For typical sample sizes and chamber geometries and without rotation, however, 40 to 60 observable oscillations are generally a good value. Since this is one to two orders of magnitude less than typical layer thicknesses in MBE, RHEED oscillations cannot in general be used to monitor complete growth sequences used for devices.

The recovery behavior of the intensity after growth is a function of the position in the monolayer growth cycle where growth was terminated [165, 166]. This means that the speed of the fast process at the onset of recovery depends on the completion of a monolayer. The island size and its inverted morphology, the hole size, reach a maximum about halfway through a monolayer growth cycle, where the islands start to coalesce and holes begin to form. This means that the mass transport required for this configuration to approach the equilibrium configuration, characterized by vanishing islands and holes, is at its largest. The relative initial slope of the recovery can therefore be used to measure relative roughness in repeated measurements.

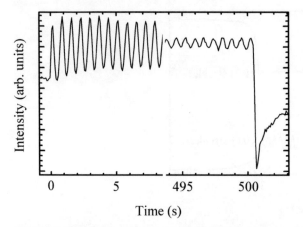

Fig. 3.4. RHEED oscillations with very low damping: $Al_xGa_{1-x}As$ ($x = 0.5$) grown on GaAs at about 20 °C below the optimum for GaAs $\beta(2\times4)$. After 720 oscillations, the amplitude is still about 1/6 of the starting value

Whereas the results presented so far suggest a correlation between surface roughness and RHEED intensity, this notion is no longer supported when we look at the phase of the oscillations. For repeated measurements with the same growth conditions, the phase of the oscillations depends strongly on incidence angle and azimuth [167,168]. We cannot, therefore, generally identify any feature in the RHEED oscillation cycle with a particular position

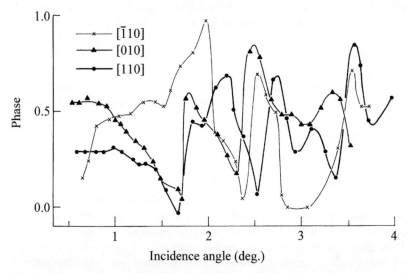

Fig. 3.5. The phase of RHEED oscillations as a function of incidence angle and azimuth for GaAs (2×4) and 12.5 keV electrons. The phase is defined by the position of the second minimum, measured from growth initiation [167]

in the layer growth cycle. Although the period of the oscillations corresponds very accurately to the period of the monolayer deposition, the phase is a complicated function of the diffraction conditions. The drastic variation of the phase with incidence angle is shown in Fig. 3.5 for the GaAs (2×4) reconstruction.

Moreover, the RHEED oscillation phase also depends on the growth conditions. Figure 3.6 shows data adapted from Briones et al. [169]. The phase in this experiment is a linear function of the As pressure within the range of each surface reconstruction. Different surface reconstructions are characterized by different constant slopes of the phase-shift-versus-As$_4$-pressure relationship.

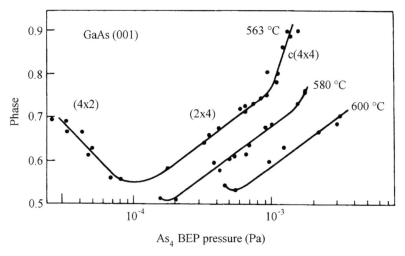

Fig. 3.6. Phase of RHEED oscillations as a function of As$_4$ pressure for GaAs (001) homoepitaxy measured along the [110] azimuth, [169]. The growth rate was 120 nm h^{-1}

At low incidence angles, extra RHEED oscillation maxima are observed for some materials [170]. This phenomenon can be identified as a diffraction effect, since these extra maxima are not present at other incidence angles. Additional maxima can also be seen at higher incidence angles, close to the values of rapid phase change in Fig. 3.5 [167]. Even if only one maximum and one minimum per ML growth cycle are present, the Fourier spectrum of the oscillations often contains strong higher-order components, depending on the diffraction conditions [171].

Apart from the intensity, the FWHM of the diffraction spots [172, 173], as well as their mutual separation [174, 175], can exhibit oscillatory variations during layer-by-layer growth. An example is shown in Fig. 3.7, where the FWHM of the specular reflection and the separation of the first-order streaks are plotted together with the intensity oscillations of the (01) streak at several points in a growth sequence. The curves are not always in phase, suggesting

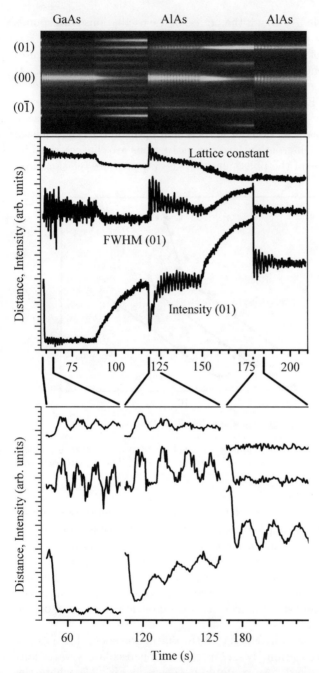

Fig. 3.7. Lattice constant, FWHM and intensity measured on the (01) streak as a function of time during epitaxy. A growth sequence including homoepitaxy of both GaAs and AlAs, as well as the heterointerface, is shown. The intensity-oscillation phase is independent of the phase of the two other oscillations

that the different quantities characterize different processes. In contrast to the intensity, the FWHM of the reflections does not seem to show any phase dispersion with incidence angle. This agrees with the assumption [86] that a kinematical analysis is valid in the directions parallel to the surface. The varying FWHM is then a measure of the lateral island-size distribution, which is purely growth-condition-dependent and should not be affected by a change in diffraction conditions.

The lateral spacing of the streaks is a direct measure of the surface lattice constant. Oscillations of this quantity have been linked to strain relaxation within the ML-high islands during layer-by-layer growth [174] in strained heteroepitaxy. As Fig. 3.7 demonstrates, they are also present in homoepitaxy. Since the amplitude of the oscillations is small, this does not necessary mean that the surface lattice relaxes in the growth islands. It could also be due to variations in the background intensity, causing only an apparent shift of the peak positions.

Summarizing the experimental results on the phase of RHEED oscillations, we are forced to conclude that a direct connection between surface roughness and RHEED intensity is insufficient to explain the observed phenomena. Instead, the variations of surface roughness and the RHEED intensity variations are both consequences of layer-by-layer growth, and any proportionality between the two quantities in general is, in general, merely coincidental.

3.2 Theoretical Models

Since a fairly rigorous treatment of the static RHEED case, even for small unit cells, is at the limit of current numerical computing capacity, state-of-the-art models for RHEED intensity oscillations involve a considerable amount of simplification. This does not have to be a disadvantage, as we shall see when we investigate basic models in Chap. 9. A simple model that qualitatively describes the basic mechanism can be more useful in designing an experiment, as well as in interpreting trends or isolated features of data, than an accurate theory that is too elaborate for a quick estimate.

3.2.1 Birth–Death Models

Some of the processes involved in layer-by-layer growth can be described by a basic rate equation introduced by Cohen et al. [122]. This so-called birth–death model describes the growth process in terms of the coverages $0 \leq \Theta_n \leq 1$ of the n molecular layers involved in the growth process. The time evolution of the system is described by the following set of differential equations:

$$\frac{\mathrm{d}\Theta_n}{\mathrm{d}t} = \frac{1}{\tau}(\Theta_{n-1} - \Theta_n)$$
$$+ D(\Theta_{n+1} - \Theta_{n+2})(\Theta_{n-1} - \Theta_n)$$
$$- D(\Theta_n - \Theta_{n+1})(\Theta_{n-2} - \Theta_{n-1}). \tag{3.2}$$

New material is added in each layer in proportion to its exposed area. Material transport takes place only downward, in proportion to the unburied area of the upper level and the empty area of the lower level. The growth rate is $1/\tau \, \mathrm{ML\,s^{-1}}$ and the amount of interlayer diffusion is governed by the diffusion parameter D. This simple model already describes the most basic features of the layer-by-layer growth process. The total roughness can be computed by assuming a parabolic roughness function for each layer, like the one shown in Fig. 3.9. This is a reasonable approximation for a layer that is smooth at $\Theta = 0$ and $\Theta = 1$ and has a roughness maximum somewhere in between. For a small diffusion parameter D, the total roughness becomes large since numerous layers are occupied at a given time. The resulting oscillations are

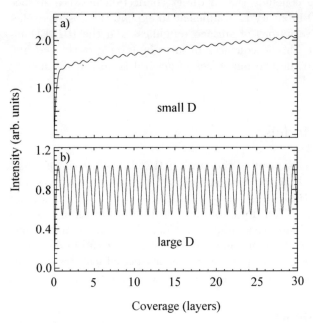

Fig. 3.8a,b. Simulation of surface roughness during layer-by-layer growth using the birth–death model. The amplitude of the oscillations is larger for a large diffusion parameter D

given in Figs. 3.8a and b for small and large diffusion, respectively. As expected, the amplitude of the oscillations is smaller in Fig. 3.8a, for a spread of the growth front over many monolayers, than in Fig. 3.8b, which corresponds to almost perfect layer-by-layer growth. A phase shift is observed for small diffusion, which is associated with the numerous small contributions of layers with coverages close to either 1 or 0. This also causes the monotonic increase of the curve in Fig. 3.8a. This effect is not very realistic as it would correspond to either needle-like protrusions or narrow holes in the surface for the usually assumed solid-on-solid model, where no overhangs or vacancies are allowed.

The model is also not very accurate in modeling the damping of the oscillations, since it does not describe the distribution of the atoms within any given layer and therefore contains very limited information about the surface morphology. It shows an approach towards a steady-state roughness, but the transition is too fast compared to realistic experiments.

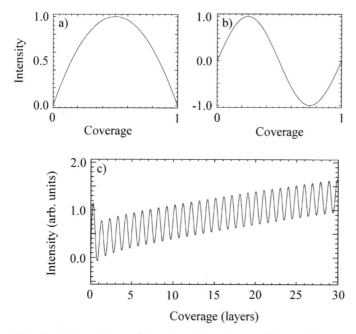

Fig. 3.9. A roughness (**a**) and an interference (**b**) model to describe the relation between the observed signal and the coverage of a layer; (**c**) shows the result using model (**b**) with the same parameters as in Fig. 3.8a. The intensity is to be taken relative to a sufficiently positive base level

Although the roughness oscillations may look similar, at least qualitatively, to observed RHEED intensities, the birth–death model cannot be used to support any correspondence between roughness and diffracted intensity. In fact, any nonconstant dependence used to compute the signal produces qualitatively similar results. This is demonstrated in Figs. 3.9b, c using a sine function. A dependence of this type could be caused by an interference effect. The curve in Fig. 3.9c uses the same diffusion parameter as in Fig. 3.8a, with a similar result. It also demonstrates that the monotonic increase of the calculated curve is due to the layers with low coverage ('needles' instead of 'pinholes'). For these layers, both coverage terms are small, leading to a quadratically small transfer rate to the lower layer. Combined with the linear rate of material addition from the first term of (3.2), this leads to the observed low coverage buildup.

3.2.2 Kinematical Model

Two basic phenomenological approaches to explaining RHEED oscillations have dominated the discussion during the last few years. One of them is a kinematical treatment that treats diffraction mainly as a single scattering event, emphasizing the interference of beams from different terraces of the surface [114,176]. This approach can explain several effects associated with diffraction from static vicinal surfaces, such as the spot splitting along the streaks [177,178] (see also Sect. 5.2). However, it fails to describe many other phenomena associated with RHEED oscillations. No oscillations should be present for the in-phase condition, where there is constructive interference of the upper and lower levels of the growing layers, but in experiments, strong oscillations can often be observed at the corresponding angles. The model also cannot account for any phase shifts, since the oscillation maxima always occur at integer coverages.

3.2.3 Edge-Scattering Model

In the second approach [130,179–182], RHEED oscillations are assumed to be composed of two contributions. One is the coherent Bragg diffraction of the kinematical approach. This involves atoms at positions that define the long-range periodicity of the surface. The second contribution is defined as the incoherent scattering proportional to the density of step-edge atoms or other defects that vary with the evolving surface morphology [153,183]. At out-of-phase conditions between the Bragg reflections, the diffracted intensity should be dominated by the destructive interference of the kinematical model, leading to minima at half-layer coverage. At in-phase or Bragg conditions, the coherent process should not produce any intensity variations, because the contributions from both levels are in phase. Instead, the incoherent process should dominate here, producing an intensity maximum at half-layer coverage, where the step density is largest. The observed phase variation could then be explained by continuous variation between these two extreme cases, alternating between a dominance of coherent and incoherent contributions. This phenomenological approach can explain phase shifts, as well as strong oscillations at the in-phase angles. Owing to the unclear definition of 'coherent' and 'incoherent' processes, however, it is not easily quantified [184, 185]. Studies based on the evaluation of STM images find a direct correspondence between step density and RHEED intensity for special cases during growth [183,186]. In these experiments, the RHEED intensity is claimed to be measured at in-phase conditions (the incidence angles are not reported) to eliminate the kinematical part of the scattering in the edge-scattering model. This signal is compared to the sample surfaces obtained after quenching the sample at different positions in the growth cycle. Usually, good agreement between the step density and RHEED intensity is found [187]. However, no systematic variations of the diffraction conditions were performed and the precise scattering geometry is not reported. Therefore, these results do not allow a more detailed investigation of the edge-scattering model, since with a suitable choice of diffraction conditions, any phase position can be obtained, as is evident from the experiments presented in Sect. 9.2.2.

3.2.4 Dynamical Approaches

Since dynamical simulations of static surfaces are already quite involved (see Sect. 2.3.2), the application of the current theoretical models to a realistic surface configuration during growth does not seem feasible at this time. One possibility of simplification, however, is to neglect all lateral modulation of the potential in the directions parallel to the surface and retain only the z dependence. Calculations of this type [188–191] use the birth–death model to determine layer coverages, which then serve as scaling factors for the potentials of the layers. The potential of each layer is identical in shape to the bulk potential. The results show a strong variation of the oscillation phase as a function of incidence angle. Quantitative comparisons with experiments, however, were not made. Since the model does not contain any lateral modulation of the potential, it cannot describe any azimuthal effects such as, for example, the phase shift for angles off the high-symmetry direction [168]. Several efforts are currently under way to include disorder in the framework of rigorous dynamical theories in a more detailed way. This disorder includes atom vibrations and point defects [143], as well as configurational disorder [147], mainly along the beam [144–146]. These models are not yet at a stage where realistic RHEED oscillation experiments could be modeled, but the rapid progress in this field allows the expectation that in the near future this gap may be bridged.

3.2.5 Top-Layer Interference Model

Inspired by the parallel slicing method, Horio and Ichimiya [192] found a simple approximation for a dynamical scattering model of RHEED oscillations. Whereas in the parallel-slicing method, the potential within each ML perpendicular to the surface is divided into several sublayers, these authors assume a constant potential within each whole ML. As in any model neglecting lateral structure, a low-symmetry direction for the incident azimuth is assumed, the so-called one-beam condition. Owing to the long-period translational symmetry perpendicular to the beam in this case, the lateral potential modulation is low and can be neglected. This leads to a model where the nth ML from the top of the crystal is treated as a homogeneous slab that is described solely by the value of its constant potential, V_n. For a nonreconstructed surface, the potential of a growing layer is then proportional to its coverage Θ:

$$V_n = \Theta_n V_0 \,, \tag{3.3}$$

where V_0 is the bulk value of the average crystal potential. The determination of the boundary conditions, i.e. the continuity of the wave function and its derivative at the boundaries, then yields the specular spot intensity.

Horio and Ichimiya performed this calculation for a multilayer system with damping, where the individual coverages were determined by a birth–death model and the damping was included as an imaginary part in the potential. They were able to reproduce the main features of the more complicated parallel-slicing transfer matrix calculation. Like the more complicated models, this model also contains phase shifts of the RHEED oscillations with respect to layer completion, as well as the phenomenon of additional maxima at low angles of incidence.

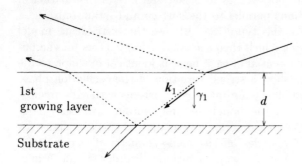

Fig. 3.10. Simulation of the specular spot intensity by the interference between the electron beams reflected from the top of a growing layer and the interface between the growing layer and the substrate. Adapted from Horio and Ichimiya [192]

In its most elementary form, the top-layer interference model neglects damping and includes only one growing layer, corresponding to the birth–death model in the limit of a very large surface diffusion parameter. The beam geometry for this case is given in Fig. 3.10. Since only the surface-normal components of $\boldsymbol{k}_0$ and $\boldsymbol{k}'$ are affected by the potential step, the phase difference $\Delta\varphi$ for the specular beam is given by

$$\Delta\varphi = 2\gamma_1 d \,, \tag{3.4}$$

where d is the thickness of the growing layer and γ_1 is the magnitude of the normal component of $\boldsymbol{k}_1$ and $\boldsymbol{k}'_1$, the incident and diffracted wavevectors inside the crystal. The boundary conditions of the wave functions give

$$\gamma_1 = \sqrt{k_\perp^2 + \frac{2me}{\hbar^2} V_1} \,, \tag{3.5}$$

where $k_\perp$ is the normal component of the wavevectors in vacuum. In Fig. 3.11, $\Delta\varphi$ is plotted as a function of d and V_1 for an incidence angle of $1°$ and an elec-

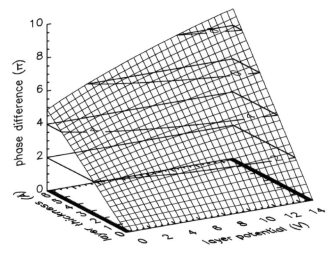

Fig. 3.11. The phase difference $\Delta\varphi$ of the top-layer interference model as a function of layer thickness d and crystal potential V_1 for an incidence angle of $1°$

tron energy of 20 keV. The left-hand edge of the diagram, $V_1 = 0$, corresponds to the kinematical case, with a linear dependence of $\Delta\varphi$ on d. Increasing the potential to the right, one can clearly see that the phase difference due to the top-layer potential can assume large values. Dynamical calculations for GaAs generally use bulk potential values between 13 and 14 V [106]. Using this value for V_1, the top-layer interference addition to the kinematical phase difference amounts to more than 2π for a layer thickness d of 1 ML (2.83 Å). The phase difference $\Delta\varphi$ shows a strong dependence on the incidence angle. The deviations from the kinematical case, $V_n = 0$, are most pronounced at small angles. This is shown in Fig. 3.12. Whereas increasing the potential from 0 to 15 V at an angle of $3°$ causes an additional shift of only about π, this shift is more than 3π for incidence angles close to zero.

In both Fig. 3.11 and Fig. 3.12, the in-phase conditions $n2\pi$ (n integer) are marked by contour lines. During the growth of one ML, the potential of the top-layer grows linearly from zero to its maximum value. At this maximum, the potential difference between the layer and the bulk is zero, the bottom interface of the layer becomes transparent, and the reflectivity is back to its starting value. Local maxima in the reflectivity occur whenever a line along the crystal potential axis of Fig. 3.12 crosses the in-phase condition. Since this can happen at any coverage, the model explains the arbitrary phase position of RHEED oscillations as a function of incidence angle as well as the variation of this phase with changing incidence angle. At small incidence angles, several in-phase conditions are met during the growth of one ML, explaining the phenomenon of multiple maxima.

The top-layer interference model can explain these phenomena with a single, simple concept. As a two-beam model it is the lowest-order approx-

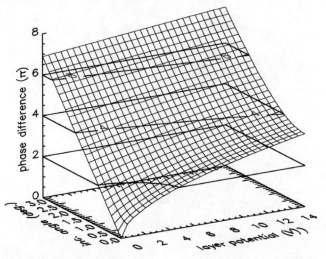

Fig. 3.12. Phase difference of the two beams in the top-layer interference model plotted as a function of incidence angle and layer potential

imation to dynamical theory. It can be incorporated in kinematical models as a correction without destroying the computational economy of the calculations. We shall therefore employ it in this work to interpret RHEED data from static as well as growing surfaces.

4. Semikinematical Simulations of RHEED Patterns

Kinematical calculations are very economical with respect to computer resources. They basically involve a summation of phase factors without any iterations. The calculation time therefore depends linearly on the number of scatterers and allows the treatment of large unit cells. In this way, even surface superstructures of complicated subunits can be simulated with relatively little effort. From the experimentalist's viewpoint, a rough and quick estimate of a diffraction pattern is often more useful than a comprehensive, but involved treatment. Kinematical theory is ideal for this purpose. The typical problems associated with RHEED comprise the determination of material parameters such as atomic spacing, surface roughness and crystal potential, or the test of proposed surface structures. If kinematical theory is sufficiently accurate for these purposes, its efficiency makes it a versatile tool for the experimentalist.

In this section we want to test the ability of kinematical theory to distinguish between different proposed reconstruction models for a given RHEED pattern. After presenting the mathematical model used in the simulations, we compare calculated reciprocal-space maps of the three models with the experimentally found pictures. This in turn leads us to the investigation of supercells and domain structures in the last section of the chapter.

To adapt the general kinematical model of Sect. 2.3.1 to the RHEED geometry, the basically two-dimensional nature of RHEED needs to be taken into account. Since the reciprocal lattice of a 2D surface is a regular arrangement of 1D rods perpendicular to the sample surface, the discreteness of the reciprocal lattice can only be maintained in the surface plane. The scattering amplitude (2.19) then becomes

$$F(\boldsymbol{G}_{\|}, \gamma, s) = A \sum_i u_i(s) \exp(-B_i s^2) \exp\left[i(\boldsymbol{G}_{\|} \cdot \boldsymbol{r}_{i\|} + \gamma r_{i\perp})\right] , \qquad (4.1)$$

where the subscripts $\|$ and $\perp$ denote the surface-parallel and surface-normal components of the respective vectors, and $\boldsymbol{r}$ and $\boldsymbol{G}$ are the discrete real-space and reciprocal space vectors. The continuous reciprocal-space coordinate g determines the position along the rod normal to the surface. As long as only relative intensities are considered, the absolute value of the constant A is unimportant and it can be used to normalize the resulting intensities.

This treatment still assumes perfect infinite periodicity along the surface, so that the calculated intensity distribution in reciprocal space consists of truly one-dimensional rods with zero width. To display the results so that they can be compared to an experiment, the rods have to be broadened artificially, but within the framework of the model this width has no physical significance. Since the rods, as well as the Ewald sphere, are broadened in the experiment, their intersection is a two-dimensional pattern instead of the row of points along the Laue circle produced by the strict theoretical treatment. For small incidence angles, this broad intersection region is a good approximation to the reciprocal-lattice plane perpendicular to the beam direction and the sample surface, since the Ewald sphere and the plane intersect at a very small angle. In these cases, the whole central region of the calculated planar intensity distribution can be used in the fitting procedure. This allows a much better comparison between theory and experiment since the information content of a two-dimensional pattern is higher. One has to keep in mind, though, that even at low angles the intensities closer to the Laue circle are enhanced in an experiment.

In the following, we begin by discussing the GaAs (113)A surface because it allows a simpler theoretical treatment, and then continue with the GaAs (001) $\beta(2\times4)$ surface reconstruction.

4.1 Different Models for the Surface Reconstruction of GaAs (113)A

Figure 4.1 shows a RHEED pattern of a GaAs (113)A surface looking along the $[\overline{3}\overline{3}2]$ azimuth. Optimum growth conditions for GaAs (113)A were used, which were found to be at a slightly lower As pressure and about $25\,°$ higher sample temperature than the optimum for the GaAs (001) $\beta(2\times4)$ reconstruction. Additional details can be found in Sect. 10.1.2. For comparison with the simulations, the length marker of the simulations corresponding to the reciprocal-lattice constant of bulk GaAs is included in the photograph.

Our starting point for comparing the two (113)A surface models is unrelaxed structures including only the surface As and Ga atoms at bulk lattice positions. This means that for the Nötzel model, possible rearrangement and reconstruction of the facets themselves [193,194] are not taken into account. For the Wassermeier model, the As dimers are not allowed to relax. The Doyle–Turner and the Debye–Waller factors in (4.1) are set to unity at this stage. Quasi-three-dimensional representations of both structures are shown in Fig. 1.5. The simulated structures and the resulting patterns for both models are shown in Fig. 4.2. The length of the marker is $(2\pi/5.6533)\,\text{Å}^{-1}$, the reciprocal-lattice constant of GaAs; diffraction intensities are represented by the gray scale given at the bottom edge of each simulation. In the following, we will refer to the model by Nötzel et al. as model A and to the Wassermeier

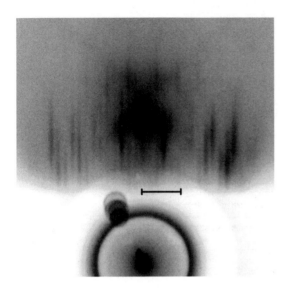

Fig. 4.1. RHEED pattern of the GaAs (113)A surface reconstruction. The observation direction is along the [$\bar{3}\bar{3}2$] azimuth with an acceleration voltage of 20 kV and an incidence angle of 1.5(2)° at a sample temperature of 605 °C

et al. structure as model B. For comparison, a simulation of 22 planes (11 As and 11 Ga) of bulk material is given in Fig. 4.3. The incidence angle and scale of the simulations are matched to the experimental result. The simulations for models A and B include the bulk spots since they are subsets of the structure shown in Fig. 4.3.

Model A exhibits a splitting of all streaks with a period corresponding to $(2\pi/10.2)\,\text{Å}^{-1}$ as expected for a two-level system [111]. Although the facet planes are only a few lattice constants wide, emerging facet streaks perpendicular to the ($33\bar{1}$) and ($31\bar{3}$) planes begin to form as a modulation of the reconstruction streaks. None of these features are present in the experimental intensity distribution of Fig. 4.1. All streaks show a slowly varying intensity distribution normal to the surface. No clear evidence of the facet streaks can be seen. The X-shaped intensity enhancement centered about 0.5° above the specular spot is due to Kikuchi lines. Only streaks 13–18, counting from (00), resemble the simulated facet streaks, but they form a much lower angle towards the surface normal than expected for model A. Instead, the angle is very similar to the one predicted for model B. In addition, the simulation of model A predicts dark areas on both sides next to the specular spot position, as well as a strong reflection on the ninth streak close to the shadow edge. Both of these features are absent in Fig. 4.1. The model B simulation is in closer agreement with the experimental pattern. The streak modulation has about the same period, and there is some intensity adjacent to the specular beam position. Several streaks already show a staircase structure similar to streaks 8–13 in Fig. 4.2b.

Whenever more atoms are included in simulating either the Nötzel or the Wassermeier structure, the pattern approaches the one shown in Fig. 4.3. The

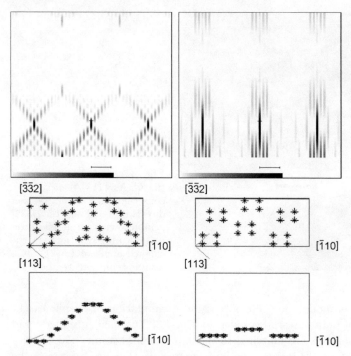

Fig. 4.2. Simulated RHEED patterns of (113)A GaAs in the $[\bar{3}\bar{3}2]$ azimuth (*left*) for the model by Nötzel et al. (model A) and (*right*) the model by Wassermeier et al. (model B). In the *bottom* part of the figure, the scatterers are plotted at their positions in the surface unit cell. The sizes of the stars are proportional to their form factors. The {001} directions of the cubic bulk unit cell are indicated by the *light-gray* axes. In the simulated patterns, the intensities are given according to the gray scale at the *bottom* of each panel. The *short horizontal lines* mark the position of the specular spot for an incidence angle of 1.52°

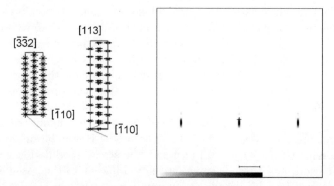

Fig. 4.3. Simulation of the RHEED pattern using 22 layers of bulk material in the (113) orientation. The bulk reflections are present in both simulations of Fig. 4.2

reconstruction features weaken and the relative intensity of the bulk reflection increases. The experimental pattern, however, shows a high intensity in the reconstruction reflections which is comparable to the intensity in the bulk streaks. Also, the modulation of the streaks along the surface normal is less in Figs. 4.2b and 4.1 than in the simulation of 22 bulk layers in Fig. 4.3. This means that the intensity distribution in Fig. 4.1 is dominated by the surface atoms of the reconstruction. Since the simulation of Fig. 4.2b is already closer to the bulk distribution than the experimental result, we conclude that fewer atoms contribute noticeably to the formation of the experimental pattern and that the surface atoms relax from their bulk positions, thus redistributing intensity to the reconstruction streaks.

In fitting the calculated pattern for model B to the observed pattern, we therefore employ two sets of fitting parameters, namely the number of scatterers and the relaxation of the surface atoms. A top view of model B is given in Fig. 4.4. The unit cell of the surface reconstruction is indicated by

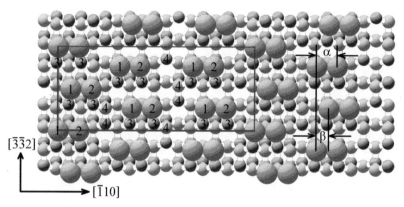

Fig. 4.4. Top view of the model B surface reconstruction of (113)A GaAs; the numbers indicate the atomic positions included in the various simulations. Possible relaxations of dimerized As atoms are described by the parameters α and β

the rectangle. The larger spheres, as in Sect. 1.3, denote the top As atoms that are assumed to form dimers analogous to the (001) surface. Since these atoms are able to relax mainly along [$\bar{1}10$], the two fitting parameters α and β are introduced to describe these displacements. The parameter α denotes the dimer bond length and β describes a possible relative displacement of the two subrows along [$\bar{3}\bar{3}2$]. For the unrelaxed structure with the top As atoms at bulk positions, $\alpha = 4\,\text{Å}$ and $\beta = 2\,\text{Å}$. The simplest approximation to the structure is to neglect the substructure of the units comprising atoms 1, 2 and 3 in Fig. 4.4 and to use only atoms 1. The resulting reciprocal-lattice image is shown in Fig. 4.5. Compared to Fig. 4.1, the result is closer to the experimentally observed pattern in that the intensity is more evenly distributed in reciprocal space. Also, the intensity variation along the individual rods is

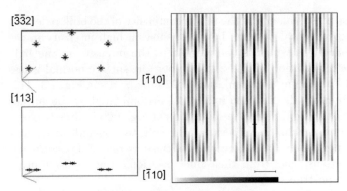

Fig. 4.5. Simulated RHEED intensity distribution of GaAs (113)A for model B using only one of the top As atoms per dimer

closer to the experiment. The calculation correctly reproduces the first three streaks of the experimental pattern, counting from (00). This characteristic structure consists of alternating regions of high intensity (h) on streaks 0 and 3 or 1 and 2 and low intensity (l) on the other streaks. The alternating h–l–l–h–l–l–h and l–h–h–l–h–h–l structure is characteristic of the (113)A GaAs reconstruction and can be found in any experimental RHEED pattern of this surface showing the eightfold periodicity along $[\bar{1}10]$.

The calculated pattern deviates from the experimental one in Fig. 4.1 on the streaks further away from (00). These streaks are more sensitive to the fine structure of the reconstruction than the ones closer to (00). This can be demonstrated by including atoms 1 and 2 in Fig. 4.4, which are the dimerized As atoms, and varying their separation α. The resulting reciprocal-space maps are shown in Fig. 4.6. One can clearly see the strong modulation of the reciprocal lattice upon varying the substructure of the surface reconstruction. The reciprocal relationship between real-space and reciprocal-space distances is nicely illustrated by the band of higher-intensity streaks moving in towards (00) from both sides as the dimer separation grows. The second-order band of higher intensities moves in from the border of the frame and is centered on the 16th streak when the dimer length reaches 4 Å. The experiment of Fig. 4.1 shows high intensity on streaks 9, 10, 12 and 13. A good approximation to this situation is reached around $\alpha = 2.7$ Å, where the intensity of these streaks is highest on average.

The results of varying both α and β are shown in Fig. 4.7. The reciprocal-space maps indicate best agreement for α around 2.7 Å and β around 0.7 Å. The value found for α agrees well with theoretical as well as experimental values for the dimer bond length [139, 195]. The subrow relaxation parameter β is quite large, even if we consider a repulsive interaction of the empty-state dangling orbitals of the As dimers. Its introduction mainly serves the purpose of demonstrating the method's ability to fit several parameters in a way that allows us to distinguish between the effects each parameter has on

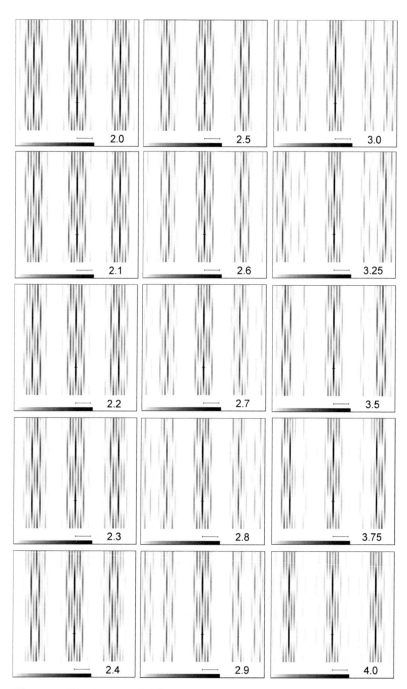

Fig. 4.6. Simulated RHEED intensity distribution of GaAs (113)A for model B using only the top As dimers. The separation of the As atoms in the dimers α is varied from half the bulk separation ($2\,\text{Å}$) to the full bulk separation ($4\,\text{Å}$)

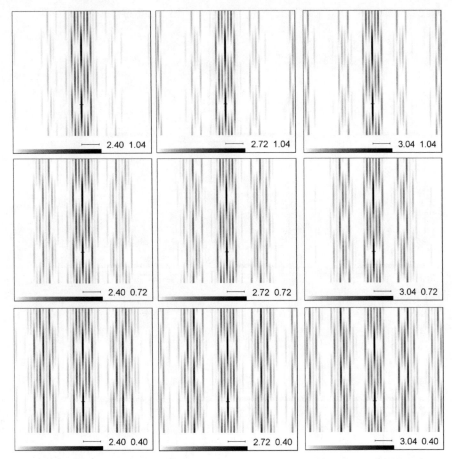

Fig. 4.7. Simulations of the (113)A GaAs RHEED pattern using model B for different values of α and β, given in units of Å by the pair of numbers below each pattern

the resulting pattern. The characteristic structure of the seven central streaks is not affected for the ranges of α and β displayed. The dense reciprocal lattice of the (113)A reconstruction allows a good fit of the relaxation parameters since any periodicity smaller than the unit cell size in real space shows up as an intensity variation on a scale larger than the streak separation in reciprocal space. The dimer separation α is about one-tenth of the real-space unit cell size along [$\bar{1}$10]. It therefore shows up as an intensity variation with a period of about ten reciprocal-lattice rods in the simulation. The same behavior holds for β, with a still larger period in reciprocal space. The accuracy of the simulation can therefore be improved by including rods further away from the (00) rod. Since the experimental pattern is the intersection of the reciprocal-lattice plane with the Ewald sphere, the truncation of the higher-order streaks limits the fitting range of the simulation. This means that, in the present case, β, because of its small value, affects the intensity only further away from the central rod, where the experimental pattern is already truncated. The parameter β is therefore a less reliable parameter in the fitting procedure than α. Comparing the central panel of Fig. 4.7 with Fig. 4.1, we find good qualitative agreement between the simulated and the experimental picture out to the 13th streak. The intensity distributions both along and normal to the shadow edge are correctly reproduced. This remarkable agreement constitutes strong evidence that the kinematical approximation is valid in this particular case.

Obviously, subsurface atoms contribute only minor corrections to the RHEED pattern. This can be explained by the smoothness of the surface along the $[\bar{3}\bar{3}2]$ direction. The low density of disorder steps along this azimuth revealed by STM [62], combined with the absence of depth modulation along $[\bar{3}\bar{3}2]$ in the surface unit cell itself, produces dense, linear rows of scatterers along the beam direction. A modulated surface that exposes areas normal to the beam direction, as in the GaAs $\beta(2\times4)$ case along [110], facilitates the entry as well as the exit of the high-energy electrons to and from the bulk. In the present case, such a modulation is not present and scattering is dominated by the surface atoms.

Streaks 4 and 12 of Fig. 4.7 do not show any modulation and do not change on variation of α or β. To obtain an intensity variation on these streaks, additional atoms such as 3 and 4 in Fig. 4.4 need to be included in the simulation. We shall discuss meaningful ways of doing this and taking into account the necessary corrections in Sect. 4.3. A discussion of the resulting patterns may be found in [196].

On the basis of the simulation results of this chapter, we can clearly confirm model B proposed by Wassermeier et al. [62] for the surface reconstruction of the GaAs (113)A surface. This is in agreement with a recent combined LEED, core-level spectroscopy and photoemission study [197]. The smaller modulation of this model is in accordance with the general observation that all known surface reconstructions involve at most the top five layers of the crystal [198]. A modulation similar to the corrugation proposed by Nötzel et al. has not been confirmed at the interfaces of $Al_xGa_{1-x}As/GaAs$ super-lattices, neither by Raman spectroscopy nor by TEM [68, 199]. More recent Raman investigations taking into account the results of Sect. 10.1.2 [200] find good agreement with an interface structure according to model B at the inverted interface, whereas the normal interface is intermixed because of segregation.

4.2 Misoriented GaAs (113)A

We have seen in the last section that a careful analysis of static RHEED patterns allows the determination of the most likely surface structure. This includes the measurement of spacings perpendicular to the substrate, if special care is taken over refraction effects (see Sect. 4.3.2). In the present section, we investigate effects due to substrate misorientation.

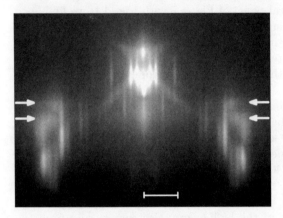

Fig. 4.8. RHEED pattern of a miscut GaAs (113)A surface along the [$\bar{3}\bar{3}2$] azimuth. Growth and diffraction conditions are similar to Fig. 4.1 except that the incidence angle is 1.2°. The sample is miscut by 0.9° towards the RHEED screen. The *arrows* indicate a Kikuchi line doublet that will be analyzed in Sect. 5.2

A typical RHEED pattern of a misoriented GaAs (113)A surface similar to the ones used by Nötzel et al. [61] is shown in Fig. 4.8. The (00) streak exhibits a triple modulation that is not present in the patterns from exactly oriented substrates, and this may lead to a misinterpretation of spacings perpendicular to the sample surface. The effect of misorientation in the kinematical picture is to tilt the main direction of the streaks with respect to the shadow edge so that they remain normal to the low-index planes. The streaks are then modulated with a period that corresponds to the step height between the terraces [114]. Since the streaks tilt away from the viewer for a miscut towards the screen, the Laue-circle center moves above the shadow edge. A complete simulation of the reciprocal-lattice distribution is therefore difficult to achieve with our semikinematical model, which produces cuts through reciprocal space perpendicular to the average surface plane, not the low-index planes. We can, however, perform a simulation perpendicular to the miscut direction to obtain a side view of the modulated streaks. These simulations were performed using a rectangular lattice to represent the symmetry of the unit cell for a 0.9° miscut similar to the experiment. Adjusting for the appropriate incidence angle, the results are combined with the experimental pattern in Fig. 4.9.

Both monolayer and bilayer steps produce maxima at the appropriate positions, providing evidence that the modulation of the central streak is a misorientation effect. Since the experiment only provides three maxima, we cannot determine whether the central maximum indicates bilayer steps or is due to the specular spot being located at the minimum of the monolayer step structure. STM experiments do not generally find bilayer steps along this direction [62].

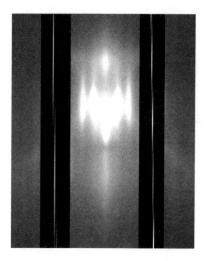

Fig. 4.9. Simulation of reciprocal-lattice rods for a miscut of 0.9° and comparison with the experimental RHEED picture. The *left insert* corresponds to monolayer steps; the insert on the *right* represents bilayer steps. For better visibility, the reader may look at the figure at grazing incidence along the rods

Another possibility is an increased modulation depth of the surface reconstruction due to misorientation. This argument assumes that a miscut in one direction also produces a change in morphology in the other direction, for example due to an elongated finger structure at the steps. This can then introduce multiple steps perpendicular to the beam direction in our example. A simple approximation for this case is a structure with twice the height of the basic structure. This configuration is simulated in Fig. 4.10. Similar struc-

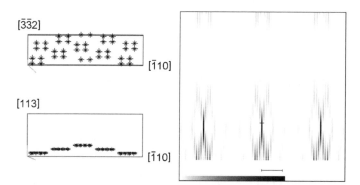

Fig. 4.10. Simulation of the (113)A GaAs RHEED pattern assuming multiple lateral steps on a structure otherwise similar to Fig. 4.4a

tures have been proposed as another possible generic surface reconstruction for the GaAs (113)A surface [201]. Since the periodicity of this structure is not twice the period of the basic structure, the lateral periodicity of the simulation does not agree with the experiment. This means that multiple steps in the [Ī10] direction, if they are present on the surface, should also produce a broadening of the rods.

At the same time, the increased depth modulation perpendicular to the surface leads to additional maxima on the central streak. The intensity profile of the central streak is extracted in Fig. 4.11 for clarity, since the modulation

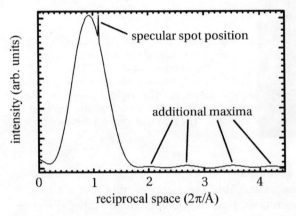

Fig. 4.11. Intensity profile of the central streak of Fig. 4.10. Additional maxima are present at higher angles

is weak. The spacing of the maxima is closer than expected from a straightforward conversion to real-space distances. This is related to the separation of Ga and As layers for the (113) orientation that adds a long-wavelength component.

From diffraction experiments alone, we cannot distinguish between the two mechanisms that could lead to the streak splitting. Both may be present in the experiment, reinforcing each other. Together, the good agreement between simulations and experiment and the absence of the modulation in the exactly oriented case indicate that the appearance of these additional maxima is due to a step structure induced by the tilting of the crystal surface.

4.3 (001) GaAs $(2\times4)/c(2\times8)$

As another application of our semikinematical model, we calculate intensity distributions for several proposed structure models of GaAs (2×4) surface reconstructions. We classify the reconstructions according to their RHEED patterns instead of their structural models. For the nomenclature, we use the convention established by Farrell and Palmstrøm [57]. To keep our discussion concise, we concentrate on the $\beta(2\times4)$ RHEED pattern and the three possible and commonly discussed structure models [202] shown in Fig. 1.3.

The aim of the following sections will be to find the most probable surface structure producing the Farrell–Palmstrøm [57] $\beta(2\times4)$ RHEED pattern. The three structure models differ in the number of As dimers in the top level and the trench configuration in the missing dimer rows, but all produce the required (2×4) periodicity. For brevity, we denote the structures in Figs. 1.3a–c as A, B and C from top to bottom. We start the discussion of the RHEED patterns with the $[\bar{1}10]$ azimuth. The $\beta(2\times4)$ structure is characterized by approximately equal intensities in the three reconstruction streaks and the fundamental streaks. From this we conclude that the contributions to the diffraction pattern from the atoms in bulk and relaxed positions are comparable. We therefore start the fitting procedure with the top layer and successively include more and more atomic layers towards the bulk. Doyle–Turner and Debye–Waller corrections will be included later, so that $u_i = 1$ and $B_i = 0$ for the time being. The top layer of all three structures consists of As dimers and is identical for structures B and C. We assume a dimer bond length of $2.6\,\text{Å}$ [195]. Using these parameters for the simulation, we obtain the patterns of Fig. 4.12. The area of the simulation is chosen so that almost

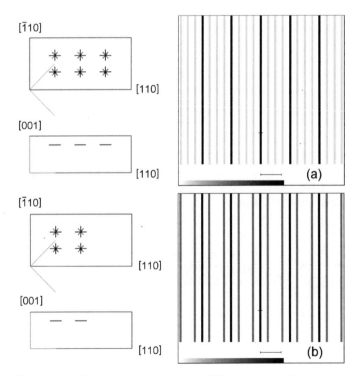

Fig. 4.12. Kinematical simulation of the reciprocal-lattice plane seen along $[\bar{1}10]$ using the top layer of the two- and three-dimer models. Panel (**a**) represents model A; panel (**b**) represents models B and C

three diffraction orders are shown. Shades of gray are used to represent the intensities; the gray scale is shown at the bottom of each figure. Black corresponds to maximum intensity. The length of the marker is $(2\pi/5.6533)\,\text{Å}^{-1}$, the reciprocal-lattice constant of GaAs. The specular beam position is marked by a cross on the central streak.

The three-dimer structure produces equal intensity on all reconstruction streaks, while the $(0\frac{1}{2})$ streak is extinct in the two-dimer models B and C (Fig. 4.12b). However, the requirement of similar intensity on reconstructed and fundamental streaks is matched better by B/C, which produces about four times the intensity on the fractional-order streaks compared to A.

4.3.1 Depth Modulation

When we add the topmost Ga layer at bulk positions to the three-dimer model A, the intensity distribution in the fractional-order streaks remains unchanged, since the second layer is not reconstructed and therefore only introduces an intensity modulation on the fundamental streaks. The resulting patterns for the two-dimer models B and C are shown in Fig. 4.13. Since the rods are modulated now, a realistic incidence angle is used. With the inclusion of the second-layer Ga, the half-order streak appears in model C. This model also exhibits modulated $\{0\frac{1}{4}\}$ streaks in accordance with experiment, whereas in model B the $\{0\frac{1}{2}\}$ streaks remain extinct. Also, the integrated intensities of the reconstruction streaks are higher for C. Model C is therefore the most likely candidate to produce the $\beta(2\times4)$ pattern. It is also evident that a dense reciprocal lattice is advantageous for fitting since it contains more streaks and therefore is more sensitive to changes of parameters.

For comparison of the experimental results with theory, we use an azimuthally integrated RHEED scan, shown in Fig. 4.14, which scans a larger portion of reciprocal space than a static pattern. For the acquisition of this pattern, the shutter of the camera was left open during sample rotation over a range of several degrees away from the $[\bar{1}10]$ azimuth, so that the moving Laue circle illuminated extended areas of reciprocal space. Since the Laue circle always includes the specular spot, intensities close to the specular spot are overemphasized. Except for the specular spot, no intensities on the central streak are represented. Keeping this in mind, we can find a best fit by qualitatively analyzing basic features of the experimental pattern and linking them to parameters of our model. Intense regions on the first-order fundamental streaks are apparent close to the shadow edge, with a smaller width along the streak compared to the other reflections. Also, reconstruction reflections directly adjacent to intense regions on the fundamental streaks are generally enhanced.

When we add further bulk layers to our simulation, the agreement with experiment deteriorates. This is demonstrated for model A in Fig. 4.15. With the two additional layers of bulk scatterers, the intensity of the fractional-order streaks is only 1/256 of the maximum intensity and disappears in the

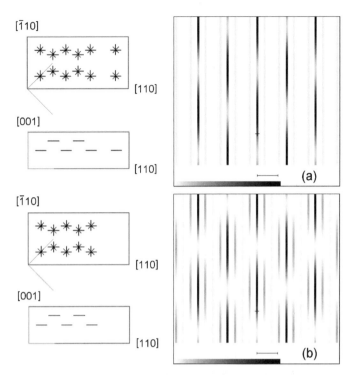

Fig. 4.13. Kinematical simulation of the (2×4) RHEED pattern using the top two layers of models B in (**a**) and C in (**b**). The incidence angle is $1.3°$

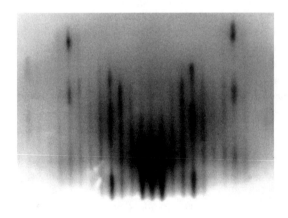

Fig. 4.14. Azimuthally integrated RHEED scan of the GaAs $\beta(2\times4)$ structure. Using a primary beam energy of 20 keV, the incidence angle was set at $1.3°$. The intensities are integrated over several degrees adjacent to the $[\bar{1}10]$ azimuth during continuous rotation of the sample around the axis perpendicular to the sample surface

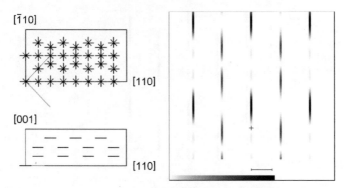

Fig. 4.15. Kinematical simulation of the reciprocal-lattice plane using the top four layers of model A (Fig. 1.3a)

display. To realize a pattern that features equal intensity on the fundamental and reconstructed streaks as in Fig. 4.14, using only the top-layer Ga and As atoms in the calculations yields better results.

4.3.2 Shadowing and Average Layer Potential

The restriction to the top layer is also justified by a geometrical argument. The RHEED patterns in this work are generally recorded along exact low-index azimuths. With the small incidence angles and significant scattering cross-sections involved, there is an efficient shadowing of lower-level atoms underneath the surface rows. At very low incidence angles of the electron beam, a row of atoms forms a dense arrangement that blocks atoms in its shadow from the incident electrons. The subsurface rows therefore contribute very little to the kinematical pattern.

The shadowing of subsurface scatterers can also explain the commonly observed fact that RHEED intensities are generally weak when viewed along low-index azimuths. Along these azimuths, an ordered surface exposes gaps between the atomic rows to the incident electrons that allow efficient channeling into the depth of the material. The channeled electrons, if not captured in the bulk, generally contribute to the Kikuchi pattern or the background and are therefore missing in the intense features of the pattern. If, however, the sample is slightly rotated, the channeling gaps are closed. Then, the surface cross-section as viewed from the direction of the incident electrons is greatly increased and the pattern becomes brighter. The electrons that channel into the bulk at the exact azimuth now contribute directly to the RHEED pattern. This shadowing mechanism enhances the agreement with kinematical theory, since it favors the scattering from top-level atoms and therefore single scattering events.

Reversing the argument, the kinematical approximation should work best for surfaces and azimuths that form straight rows of surface atoms along the beam direction. We have already demonstrated that this is the case for the (113)A surface (Sect. 4.1). It should also apply for the three GaAs models along $[\bar{1}10]$ (see Fig. 1.3).

Since we neglect multiple scattering events except for the zeroth-order Fourier component of the potential, we have to reduce the form factor of scatterers according to the probability of subsequent scattering events. A scatterer in a dense environment has a small probability of producing single scattering events. The deeper inside the crystal, the smaller the probability for kinematical scattering. As a reasonable approximation, we can assume a kinematical form factor proportional to the range of exit angles towards the vacuum. An electron scattered from a lower-level atom is shadowed from undisturbed return to the vacuum by its upper-level neighbors. A more detailed discussion can be found in [203].

Variations of the lower-level atomic form factors, however, do not drastically affect the intensities on integral-order streaks. To explain the intensity distribution on these streaks, we need to include zeroth-order dynamical scattering in our model.

A straightforward extension of (4.1) towards dynamical theory can be achieved by including the refractive effect of the average crystal potential [204], similarly to the model described in Sect. 3.2.5. Since the potential step at the crystal surface affects only the surface-normal component of the incident and diffracted wavevectors, it can be included as a single additive contribution in the phase factor of (4.1). The phase difference $\Delta\varphi$ due to refraction at the vacuum–solid interface for different incidence and exit angles is given by

$$\Delta\varphi = d_i \left(\sqrt{k_{0\perp}{}^2 + \frac{2me}{\hbar^2}V_1} + \sqrt{k'_\perp{}^2 + \frac{2me}{\hbar^2}V_1} \right) , \tag{4.2}$$

where $k_{0\perp}$ and $k'_\perp$ are the surface-normal components of $\boldsymbol{k}_0$ and $\boldsymbol{k}'$, and d_i is the real-space distance from the ith scatterer to the surface. With this dynamical correction, (4.1) becomes

$$\begin{aligned} &F(\boldsymbol{G}_\|, \gamma, s) \\ &= A \sum_i u_i(s) \exp(-B_i s^2) \exp\left[\mathrm{i}(\boldsymbol{G}_\| \cdot \boldsymbol{r}_{i\|} + \gamma r_{i\perp} - \gamma d_i + \Delta\varphi) \right] . \end{aligned} \tag{4.3}$$

The decay of the average crystal potential towards the surface is modeled as a step function at the positions of the atomic layers. This is only approximately true, since the potential at the plane containing the centers of the topmost atoms must be assumed to be finite. Therefore, the potential difference between this layer and the ones beneath it should be less than the difference between the average bulk crystal potential and the vacuum level. Nevertheless, our model should be reasonably accurate since we apply it to layers that differ in structure from the underlying bulk material. These systems can be

approximated relatively well by two rather well-defined boundaries, namely
the substrate–layer and layer–vacuum interfaces.

The modulation of the integer-order streaks in our simulations strongly
depends on the value of the top-layer potential V_1 in (4.2) between the subsurface atoms and the top atomic plane. It can therefore be used to obtain
a best fit for the V_1 value. Simulations for three different values of the potential and reduced form factors for second- and third-level atoms are given
in Fig. 4.16. For a two-level system, in which all scatterers are arranged in

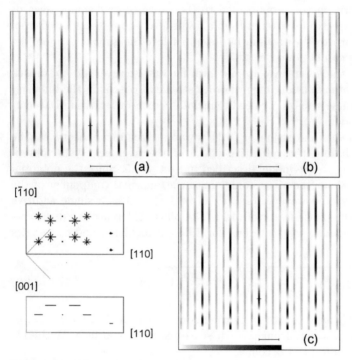

Fig. 4.16. Effect of various values of the average layer potential V_1 on the simulation of the RHEED pattern using model C. (**a**) $V_1 = 7\,\mathrm{V}$, (**b**) $V_1 = 10\,\mathrm{V}$, (**c**)
$V_1 = 14\,\mathrm{V}$

two levels, the effect of increasing the potential is to push in additional maxima from the shadow edge while compressing the diffraction pattern along
the streaks. Since the third-level atoms contribute only weakly in our model,
this model is essentially such a two-level system. The intensity distribution
parallel to the shadow edge is only weakly affected for a two-level system and
therefore the top-layer potential can be adjusted as a largely independent
parameter. When the potential is included, the modulation period along the
streaks decreases towards the shadow edge, which can explain the relatively
sharp maxima on the first-order streaks close to the shadow edge in Fig. 1.3.

The best fit with the experiment, if we adjust the simulations to these shadow-edge maxima, is achieved around $V_1 = 7\,\mathrm{V}$. Using this value, we also obtain a better fit for the overall intensity distribution on the reconstruction streaks. In particular, the strong intensities on the quarter-order streaks close to the specular spot are now at the correct position. This indicates that a mean-inner-potential correction to the kinematical model of approximately $7\,\mathrm{V}$ is appropriate. This value is only about half of the generally accepted bulk mean potential. However, the crystal bonds are relaxed at the surface. This means that the top atoms of the reconstructed surface are less strongly bound than the bulk atoms. Since the bond energies correspond to the potential, we can safely assume that the mean potential of the top layer is considerably less than the bulk value. In addition, the potential at the top plane position is already finite, which further reduces the simulated V_1.

4.3.3 Relaxation, Doyle–Turner and Debye–Waller Corrections

So far, only the positions of the dimerized As atoms have been relaxed from their bulk coordinates. The dimerization is the most significant adjustment in the reconstruction process, but not the only one. To a lesser degree, all atomic coordinates in the top few layers are affected.

First-principles calculations indicate that in our models relaxation takes place mainly parallel to the surface, whereas the atomic positions perpendicular to the surface are only weakly affected [195]. Relaxation of second- and third-level atomic positions mainly redistributes the intensity of the reconstruction streaks between different orders [203] and is not treated in detail here.

We have not yet included the Doyle–Turner scattering potentials and the thermal effects modeled by Debye–Waller factors in (4.3). The difference in atomic number between Ga and As is small, and so is their difference in scattering factors, see Fig. 2.2. Therefore, we do not expect a significant redistribution of intensities due to scattering-factor differences. Both the Doyle–Turner and the Debye–Waller factors, however, introduce an exponential decay away from the center of the reciprocal lattice in the approximation used. The intensity distribution of a pattern including these two corrections is therefore strongly enhanced towards the (000) spot.

Figure 4.17 shows a simulation using the Doyle–Turner factors of Ga and As and a Debye–Waller factor for 880 K. It is otherwise identical to the 7 V structure of Fig. 4.16, except that the form factors of the Ga edge atoms were slightly decreased, which gives a better fit on the half-order streaks. The (00) streak was excluded since it is quite intense compared to the rest of the image and would suppress the display of lower intensities in the exponentially damped tails of the streaks. One can clearly see that, apart from the radial damping, the intensity distribution is only weakly affected by the introduction of the Doyle–Turner factors. Since the center of the Laue circle does not coincide with the center of the reciprocal lattice, intensities away

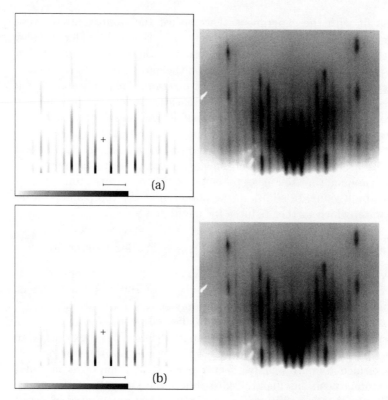

Fig. 4.17. Kinematical simulations of the RHEED patterns of model C, including (**a**) the Doyle–Turner scattering factors for Ga and As and (**b**) additional Debye–Waller corrections for 880 K. The intense central streak is excluded for better visibility of the low-intensity streaks. For each simulation, the experimental pattern of Fig. 4.14 is included for reference. The simulation in (**a**) represents the best fit to the experiment

from (00) on the Laue circle are enhanced. Compared with the experiment, the radial damping appears too strong in the simulation. This confirms that the approximated Doyle–Turner potentials overemphasize the forward direction, as already shown in Fig. 2.2b. Our model is certainly too coarse to find empirical values of the scattering potentials from the fitting procedure. The next step to improve the similarity would therefore be to include approximations to the Doyle–Turner potentials that are computationally slightly more involved but more accurate [205], and then again to compare the results with the experiments.

One effect of the strong decay of intensity away from (000) is to shift the peak positions downwards along the streaks. This effect can be observed most clearly on the {01} streaks for the second reflection from the shadow edge, in comparison to Fig. 4.16b.

Since the Debye–Waller factors are of the same form as the Doyle–Turner factors and the intensities are normalized, the temperature-related corrections just introduce an additional radial damping, which can be verified by comparing Figs. 4.17a and b.

4.3.4 The [110] Azimuth

Figure 4.18 shows an equivalent pattern to Fig. 4.14 along the [110] azimuth. The growth and diffraction conditions are identical for both images. It is evident from the procedure used to construct the simulated images in the [$\bar{1}$10] azimuth that the same set of parameters will not be appropriate for the [110] direction. Since the environment perpendicular to the beam direction determines the relative form factors of the row atoms, formerly equivalent

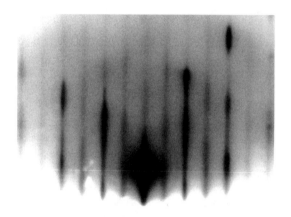

Fig. 4.18. Azimuthally integrated RHEED intensity distribution from the same surface as used in Fig. 4.14, along [110]. The incidence angle is 1.3° using 20 keV electrons

atoms now have to be weighted differently. In addition, the twofold azimuth has a widely spaced reciprocal lattice. Therefore, the fit is expected to be less reliable than for the [$\bar{1}$10] azimuth. Figure 4.19 shows a simulation using equal form factors for the second-level scatterers. The best fit is achieved for a lower-level-potential value of 8.5 V. The relative Ga form factor is 0.5.

The higher potential value can be explained by the denser environment of the Ga atoms in this azimuth compared to [$\bar{1}$10]. The steeper bond angles cause a higher electron density between the two levels, as seen from the lower-level scatterers for the low incidence and exit angles used. Compared to the experimental distribution, the second maximum from the shadow edge on the {01} streaks is displaced to lower angles. This is probably due to reduced shadowing at off-axis azimuths.

The {0$\frac{1}{2}$} streaks are not modulated in the simulation. This is due to the symmetry of the relaxation. To obtain a modulation along a reconstruction streak, the simulated structure needs to contain at least two layers that have the symmetry of the superstructure. In the present case, inclusion of

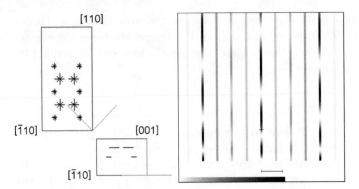

Fig. 4.19. Simulated reciprocal-lattice plane of model C along the [110] azimuth

the third-layer As dimers introduces an asymmetric modulation. This modulation cancels exactly, however, if we superimpose it on the intensity of the isomorphous structure with the third-level dimer displaced by $(-0.5, 0.5, 0.0)$ in units of the bulk unit cell. This is a consequence of the diffraction process being sensitive to correlation instead of position. A symmetric displacement of units in different layers does not induce a streak modulation unless the intrinsic symmetry of the units differs.

This can be studied by relaxing the in-plane positions of the second-level atoms. A symmetric relaxation similar to the top As dimers produces the desired streak modulation. Again, the fractional-order streak modulation and the fundamental streak modulation are independent of each other. Whereas the former is mainly related to the magnitude of the form factors and in-plane relaxation of the scatterers, the latter depends predominantly on the average potential value. This greatly simplifies the fitting procedure, since both parameters can be optimized independently.

The two isomorphic positions of the third-layer As dimers lead us to the more general problem of supercell arrangements and defects in the periodicity of the surface unit cells. So far, we have only considered a periodic arrangement of unit cells with the period of the unit cell of the reconstruction. We can, however, displace rows of the surface tiling by primitive unit cell vectors of the underlying bulk lattice to obtain supercells with periodicities much larger than in the (2×4) of the present case.

4.4 Domains

The most frequently seen superstructure of the $\beta(2\times4)$ surface is the $c(2\times8)$, where successive (2×4) rows along [110] are displaced by half the unit cell along [$\bar{1}$10]. This structure and the corresponding simulated pattern are shown in Fig. 4.20. Second-layer relaxation is included. As expected for a centered structure, the half-order streaks are extinct. The observed structure

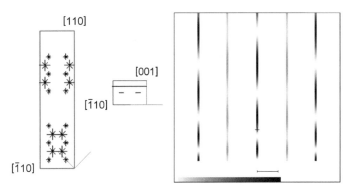

Fig. 4.20. Simulated RHEED pattern of the twofold supercell corresponding to the $c(2 \times 8)$ structure. The intense (00) streak is excluded

therefore cannot be a pure $c(2 \times 8)$ reconstruction, since the half-order streak is present in the experimental pattern, although weaker than in the fourfold direction. STM scans of the $\beta(2 \times 4)$ surface indicate roughly similar abundances of (2×8) and $c(2 \times 8)$ cells. An example is reproduced in Fig. 4.21. The

Fig. 4.21. STM scan of a $\beta(2 \times 4)$ surface showing a mixture of $c(2 \times 8)$ and straight (2×4) domains. The *dashed black lines* run along [110] to visualize the shifts of the dimer rows

dashed black lines running along [110] are included to visualize the shifts. The similar abundances of both alignments indicate that the formation energies for the two unit cells differ only insignificantly. We therefore obtain a mixture of both straight and centered cells, leading to nonvanishing intensity on the half-order streaks. The simplest supercell exhibiting this mixture of shifted and unshifted subcells is simulated in Fig. 4.22.

Another effect of the supercell arrangement is to equalize the intensities of the different Laue zones. Compared to the single-cell structure, the arrangement of Fig. 4.22 produces twice the intensity on the $\{0\,2\frac{1}{2}\}$ streaks, while

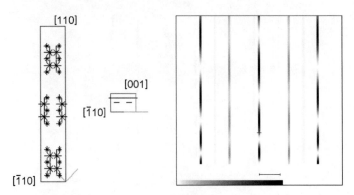

Fig. 4.22. Simulated reciprocal lattice of a threefold supercell that does not have centered symmetry

leaving the other intensities and, especially, the peak positions approximately constant. Since an arbitrary distribution of surface unit cells produces a large variety of configurations within the coherence range, we can expect this to lead to an equalization of the different Laue zone intensities. A realistic simulation would require the use of very large supercells, which was not done in the present study.

Another typical defect structure of practical importance is a kink in the double dimer rows of the $\beta(2\times4)$ reconstruction. The corresponding unit cell and simulation are given in Fig. 4.23. The unit cell shown represents the

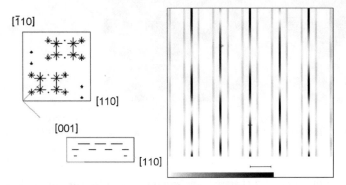

Fig. 4.23. Simulated reciprocal lattice of alternating-kink structures for model C along $[\bar{1}10]$

highest possible kink density, which exists only as the limit of real structures. Within the resolution of the display, the half-order streaks are extinct. The experimental analog of this structure is realized in the deposition of Si on the $\beta(2\times4)$ surface. This introduces kinks of the type simulated, proportional to the number of deposited dopant atoms. The appearance of the kinks, seen in

STM, corresponds to a reduction of the half-order streak intensity in RHEED similar to our simulation. We shall return to the connection between Si doping and surface reconstruction in Sect. 10.4.2.

Apart from domain boundaries within the same surface reconstruction, the case of a mixture of different surface reconstructions is also frequently observed [206, 207]. Depending on the transfer width of the RHEED instrument, the presence of surface unit cells with different symmetry leads either to an incoherent superposition of the two respective diffraction patterns or the appearance of so-called 'asymmetric' reconstructions.

Simulations for supercells composed of different numbers of single and twofold unit cell superstructures are shown in Fig. 4.24. A simple one-dimensional model is used in which each subunit consists of a scatterer and

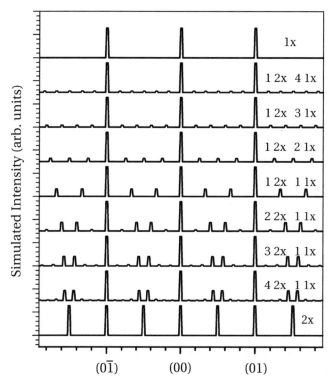

Fig. 4.24. Sections through the reciprocal-lattice rods perpendicular to the beam direction and parallel to the sample surface for supercells composed of 1× and 2× units ($k = 1$, $l = 2$)

the corresponding number of empty bulk lattice spacings. The subunits are concatenated to form a composite unit cell, which serves as the input to the kinematical simulation.

The simulated reciprocal-lattice profiles show two superstructure peaks moving in from the bulk lattice reflections and finally forming the single twofold superstructure peak at $(1\frac{1}{2})$. A similar simulation is shown for the transition from a twofold to a threefold reconstruction in Fig. 4.25. During this transition, the central reconstruction streak splits into two components

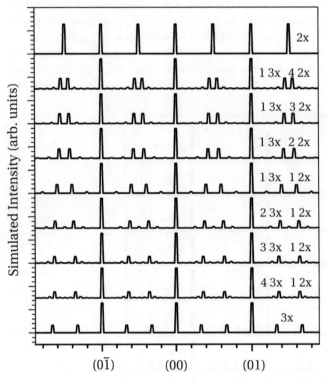

Fig. 4.25. Sections through the reciprocal-lattice rods perpendicular to the beam direction and parallel to the sample surface for supercells composed of 2× and 3× units ($k = 2$, $l = 3$)

that move outwards until they reach the symmetric positions of the threefold superstructure. Note that a given pattern does not allow the unambiguous identification of the structure. Two 2× and one 1× subunit in the sixth row of Fig. 4.24 produce practically the same result as one 3× and one 2× subunit in the fifth row of Fig. 4.25. The transition from threefold to fourfold periodicity is shown in Fig. 4.26. If we denote the superperiods of the subunits by k and l, respectively, and define the ratio r by

$$r = \frac{n}{m+n},$$
(4.4)

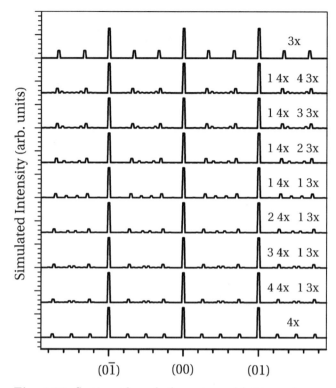

Fig. 4.26. Sections through the reciprocal-lattice rods perpendicular to the beam direction and parallel to the sample surface for supercells composed of 3× and 4× units ($k = 3$, $l = 4$)

where n is the number of k-cells and m the number of l-cells, the position x of the outermost reconstruction streak along the reciprocal-space axis is given by

$$x = \frac{1}{l + r(k - l)} ,\tag{4.5}$$

measured from (00). The value of x is normalized to the Brillouin zone size. This allows us to determine the ratio of the different subunits by solving (4.5) for r, i.e.

$$r = \frac{lx - 1}{x(l - k)} ,\tag{4.6}$$

given that we know the periodicities k and l for $r = 0$ and $r = 1$. In practice, this is only an approximation that is valid within the accuracy of the kinematical model used. For a real system, the different surface unit cells consist of more than one scatterer, each of which can have a different atomic form factor. This may lead to slight deviations from the values given here.

The asymmetric RHEED pattern due to a mixture of different subunits can also be simulated for a weighted, but random combination of subunits [208]. Simulations for $k = 2$ and $l = 3$ are shown in Fig. 4.27. The

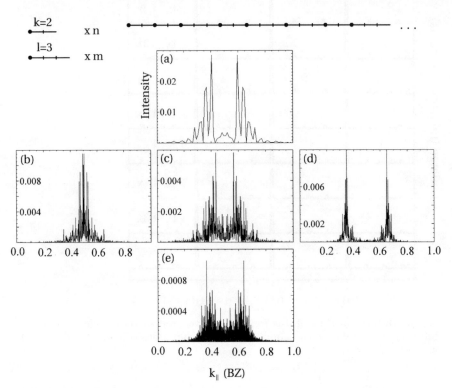

Fig. 4.27. Reciprocal-lattice profile parallel to the sample surface calculated by using a random sequence of units. With $k = 2$ and $l = 3$, the ratio r of the different units and the number of units used in the simulation are varied (*left* to *right* and *top* to *bottom*, respectively). (**a**) $r = 1$, 30 units; (**b**) $r = 0.8$, 300 units; (**c**) $r = 1$, 300 units; (**d**) $r = 0.2$, 300 units; (**e**) $r = 1$, 3000 units

real-space lattice is composed of a random sequence of the two subunits with the given ratio r. The reciprocal-space lattice parallel to the shadow edge is then again obtained by summing the kinematical phase factors. The corresponding case for threefold and fourfold reconstructions is shown in Fig. 4.28. The five panels of Figs. 4.27 and 4.28 are arranged so that the vertical direction indicates a variation in the number of scatterers included in the simulation. The horizontal direction indicates a variation of r. All simulations are normalized to the intensity of the fundamental reflections.

Note that in the sequence Fig. 4.27a,c,d, the absolute intensity strongly decreases with the inclusion of more scatterers. This is due to the infinite coherence implicit in the kinematical simulations. For large distances, the cor-

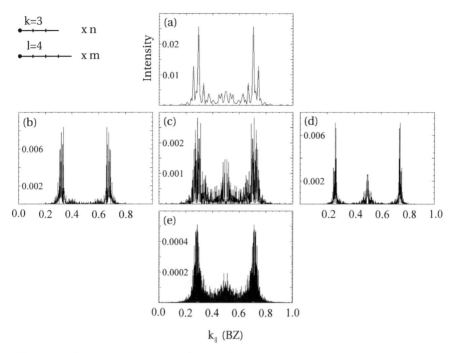

Fig. 4.28. Reciprocal-lattice profile parallel to the sample surface as in Fig. 4.27 but using $k = 3$ and $l = 4$. (**a**) $r = 1$, 30 units; (**b**) $r = 0.8$, 300 units; (**c**) $r = 1$, 300 units; (**d**) $r = 0.2$, 300 units; (**e**) $r = 1$, 3000 units

relation function goes to zero and for almost every scatterer there is another one with the opposite phase, canceling out the total signal. The variation of r (b, c, and d in Figs. 4.27 and 4.28) reproduces the peak positions obtained from the ordered simulations of Figs. 4.24–4.26. This demonstrates that the results obtained there can be generalized to a random arrangement of the two subunits. The peaks are wider around $r = 0.5$ than for arrangements closer to the pure individual reconstructions.

Experimental manifestations of asymmetric reconstructions can be found in many materials systems [206, 207, 209–214]. Both coherent superpositions leading to asymmetric reconstructions and incoherent superpositions where the two individual patterns are superimposed are observed. The realization of either one of these cases depends on the transfer width of the RHEED instrument. If the transfer width is smaller than the average domain size, the patterns of the individual reconstructions are superimposed [206], since each domain acts as an independent subsystem. If the transfer width is much larger than the domain size, all subunits coherently contribute to the same pattern, leading to an asymmetric reconstruction [207]. The superposition of individual patterns can be studied by superimposing individual simulations as shown in Fig. 4.29. In this figure, the total number of subunits is fixed at 3000.

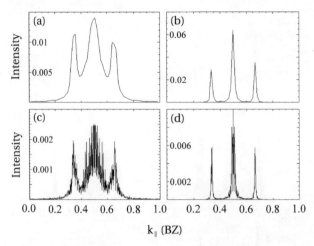

Fig. 4.29. Simulation of a finite transfer width with a constant number of scatterers (3000). In (**a**) and (**c**), $r = 1$ with 80 % pure domains. The domain size is 30 scatterers and 300 scatterers, respectively. In (**b**) and (**d**), the domains are 90 % pure, again with domain sizes of 30 (**b**) and 300 (**d**)

The ratio r of the 2× and 3× subunits is 0.5. Since the transfer width in the simulations is always equal to the total system dimension, a transfer width smaller than the total system size is approximated by superimposing the results of several independent simulations containing a high percentage of the appropriate subunits. In Figs. 4.29a,b the domain size is 30; in Figs. 4.29c,d it is 300. In Figs. 4.29a,c the domains are 80 % pure, with the remaining 20 % consisting of the other type of sub-unit; in Figs. 4.29b,d the individual domains are 95 % pure. As expected, the pattern is a superposition of the individual subunit patterns with the width of the peaks depending on the purity of the domains. Compared to the coherent superposition with the same number of scatterers in Fig. 4.27, this superposition leads to much higher intensities, since the long-range correlation of a random arrangement of subcells is absent.

The presence of an asymmetric reconstruction therefore implies the absence of domains or a domain size much smaller than the transfer width of the instrument. If the domain size is comparable to the transfer width or larger, the resulting pattern is the superposition of the two subunit patterns.

Finally, instead of leading to a new, asymmetric superstructure pattern, the presence of different domains can also cancel the effect of the corresponding substructures, resulting in a pattern that is simpler than the one that would be obtained from a single pure domain. This is the case for the (001) surface of zinc blende GaN [215], where STM scans indicate domains of a $(\sqrt{10} \times \sqrt{10})$R18.4° structure. In RHEED, a simple $c(2{\times}2)$ pattern is observed. This is due to the fact that the unit cells form domain boundaries so that the structure factors of the domains cancel. The geometry is shown in

Fig. 4.30. The [110] direction is upwards, with the rectangular grid indicating the primitive surface mesh of the face-centered cubic (001) surface. The short, vertical lines within the squares indicate Ga dimers; circles denote the positions of the missing dimers leading to the $(\sqrt{10} \times \sqrt{10})$R18.4° structure. The unit cell of this superstructure is indicated by the tilted squares. The thicker lines indicate the cubic unit cell aligned along the $\langle 001 \rangle$ directions. If we continue the dimer void positions from the left-hand domain into the

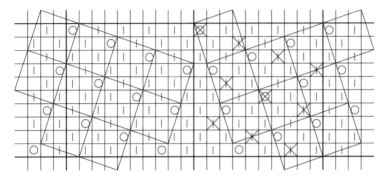

Fig. 4.30. Schematic representation of a domain boundary in the cubic GaN (001) $(\sqrt{10} \times \sqrt{10})$R18.4° surface reconstruction

right-hand domain, we obtain the positions marked ×. These coincide with Ga dimer vacancies in 20 % of the cases. The same result is obtained if the right-hand domain is shifted up or down by a multiple of two primitive surface mesh units. The 20 % probability of obtaining a dimer void position is identical to the total fraction of dimer voids. This means that the structure factors of the different domains cancel and the RHEED pattern is the same as if the surface formed a random arrangement of dimer vacancies with 20 % probability.

Quite sophisticated methods have been developed for the treatment of domain structures in X-ray diffraction [216]. The accuracy of the kinematical treatment is much better in this case and several systems have been studied [217, 218]. When applying more complicated models to the RHEED case, however, the validity of the kinematical approximation for the given problem has to be verified to make sure that the intrinsic accuracy of the kinematical model is at least as good as the expected improvement from using a more sophisticated treatment.

5. Kikuchi Lines

The present models describing the origin of Kikuchi lines are based on the assumption that the Kikuchi lines are independent of the elastic pattern. Since they are present in most RHEED patterns, they offer an additional and independent data set describing the crystal and its surface. This additional information can be used either to confirm results obtained by other methods or to obtain additional information independent of the the elastic pattern. In the following, we present a method to obtain the average crystal potential and the crystal misorientation from an analysis of the Kikuchi line pattern.

5.1 A Simple Scheme for the Geometrical Construction of Kikuchi Patterns

We start with the general scheme for describing Kikuchi lines generated in the bulk and then discuss modifications due to the crystal surface.

5.1.1 Three-Dimensional Lines

Kikuchi lines [219] can be explained by a two-step scattering model [220–224]. In the first step, the incident electrons suffer collisions that randomize the directions of their wavevectors. If the energy loss in this first collision is small compared to the electron energy, the resulting electron distribution corresponds to an electron source of (almost) the initial energy, with isotropic emission within the crystal. To obtain a clear line pattern, the energy spread of this electron source must be small. The Kikuchi line pattern is the diffraction pattern due to this isotropic electron irradiation. It can be determined using basic kinematical theory. The diffraction condition (2.1) (elastic scattering) is visualized in Fig. 5.1. Diffraction occurs for any k_0 that starts at the origin of the reciprocal lattice and ends on the plane bisecting the reciprocal-lattice vector g_{hkl}. These planes define the Brillouin zone boundaries of the crystal.

Usually, however, the diffraction geometry is treated using the Ewald sphere construction shown in dark gray in Fig. 5.1. In this case, the tip of k_0 is attached to the origin of the reciprocal lattice and its origin in turn

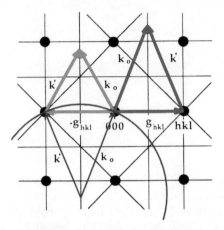

Fig. 5.1. Graphical representation of the diffraction condition (*lighter-gray vectors*) and the Ewald sphere construction corresponding to the the top-right-hand case (*dark gray*). The lighter-gray vectors end on Brillouin zone boundaries

defines the center of the Ewald sphere. The Ewald sphere construction is not very suitable for the present case of wide-angle illumination, since the construction of the k' vectors consistent with the diffraction condition involves a continuum of Ewald spheres. Instead, an approach based on the basic geometry of the light-gray vectors in Fig. 5.1 yields a more direct result. Instead of the Ewald sphere, we define a sphere of radius k_0 centered at the origin of the reciprocal lattice. This sphere is called the 'sphere of reflections' [221]. As shown in Fig. 5.2, the diffraction condition is fulfilled for the intersection of the sphere of reflections with any Brillouin zone boundary. Since the

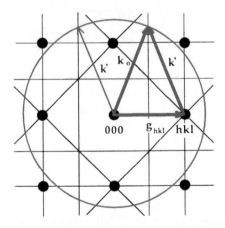

Fig. 5.2. Graphical representation of the sphere of reflections. The intersections of this sphere with the Brillouin zone boundaries determine the Kikuchi line positions

corresponding diffracted vectors k' for this condition also end at the zone boundary and start at a reciprocal-lattice point, they can be translated to the origin by a reciprocal-lattice vector, still ending on the sphere. Using this translational correspondence, a vector k' exists for any vector k_0 with the same properties, namely that it ends at the intersection of the sphere of reflections with a Brillouin zone boundary. Owing to the translational symmetry of the reciprocal lattice, the k' vectors scan the complete set of zone boundaries intersecting the sphere when k_0 assumes all possible directions.

This construction provides us with a general geometrical procedure to produce the Kikuchi line pattern, which is valid for any type and geometry of diffraction. The first step consists of determining the intersection pattern of the sphere of reflections with the Brillouin zone boundaries of the crystal. This pattern determines the possible endpoints of the diffracted vectors. The observed Kikuchi line pattern is then the projection of these intersections on the RHEED screen with the origin of the reciprocal lattice as the center of the projection.

In the RHEED case, the procedure can be simplified by taking into account the small solid angle of the screen and the large radius of the sphere of reflections. For the area visible on the phosphor screen, the intersection of the sphere with the reciprocal lattice is almost planar and parallel to the screen. Let us also assume some broadening of the range of k_0s involved in the scattering. This causes all zone boundaries not parallel to the zone axis towards the screen to be blurred sufficiently that their intensities are indistinguishable from the background. As we shall see, no lines corresponding to these nonparallel zone boundaries are observed in the experiment.

Although the generation of the Kikuchi lines is a three-dimensional process, the construction of their projection on the screen has now been reduced to a two-dimensional problem. We plot the reciprocal-lattice points in the plane parallel to the screen and construct the Brillouin zone boundaries. The geometry of the reciprocal-lattice plane can often be directly verified in the RHEED image of a rough surface that displays a transmission pattern. The resulting line pattern is shown in Fig. 5.3 for a (001) FCC crystal surface looking along the [110] direction. The Kikuchi lines are labeled according to the reflections they are derived from. In general, a few points close to the origin are sufficient to model the experimentally observed pattern, since the scattering potentials strongly favor scattering in the forward direction.

Since we assume that the pattern is generated in the bulk, we have to include in our treatment the potential correction for the beams that cross the surface from inside the crystal to the vacuum [204]. The surface-normal component $k'_\perp$ of the diffracted wavevector k' outside the crystal is then given by

$$k'_\perp = \sqrt{k_\perp^2 - \frac{2me}{\hbar^2}V}\,, \tag{5.1}$$

similarly to (3.5). The potential difference V is positive for an electron crossing the surface from the inside, leading to a reduction of its surface-perpendicular wavevector component. Only electrons with a surface-normal component $k'_\perp$ larger than the potential correction are able to leave the crystal. The lengths of the vectors are measured with respect to the shadow edge position. We therefore obtain V and the shadow edge position as fitting

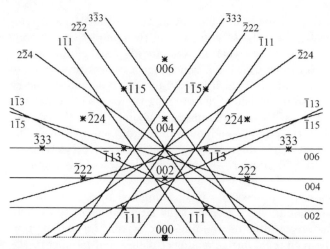

Fig. 5.3. Theoretical Kikuchi line pattern for the crystal surface geometry of an FCC (001) surface along the [110] direction. No surface-potential corrections are included

parameters from adaptation of the calculated Kikuchi line pattern to the experiment. This allows us to simultaneously obtain the misorientation angle of the substrate crystal and the average value of the crystal potential.

A fit for an exactly oriented (001) GaAs $\beta(2\times4)$ surface along $[\bar{1}10]$ is shown in Fig. 5.4. The zone axis of the Kikuchi pattern is marked by the square symbol. It does not coincide with the (000) spot of the elastic pattern, since there is no single incident-beam direction for the Kikuchi line case. The

Fig. 5.4. Simulation of the Kikuchi line pattern of an exactly oriented GaAs (001) $\beta(2\times4)$ surface in the $[\bar{1}10]$ azimuth. For line indexing see Fig. 5.3. The electron energy is 20 keV

origin of the reciprocal lattice for the Kikuchi line pattern is locked to the crystal zone axis, whereas the origin of the reciprocal lattice for the elastic pattern is locked to the transmitted beam position. Both, however, have the same orientation and geometry, as required by their construction. As expected for an exactly oriented substrate, the crystal zone axis coincides with the center of the Laue circles and is located on the shadow edge. To obtain a reliable fit, in the first step the overall size of the pattern is determined by adjusting the positions of the reciprocal-lattice points parallel to the surface so that they coincide with the corresponding streaks. The adjustment of the pattern perpendicular to the sample surface is then performed using the position of the origin and the potential value as fitting parameters. It is advantageous to take reference points far away from the shadow edge as well as features close to it, such as the ones marked by the arrows in Fig. 5.4. The 004 line in the present case is the most sensitive to the potential as it runs parallel and close to the shadow edge. The potential is then adjusted so that the relative distances of the features close to and far from the shadow edge agree with the experiment. For the simulation of Fig. 5.4, a potential value of 10.5 ± 1 V is obtained.

5.1.2 Extension to Fewer Than Three Dimensions

For two-dimensional or one-dimensional lattices, the Brillouin zone boundaries are different, therefore producing different sets of Kikuchi lines. It is important to note in this context that the reciprocal lattices of 2D and 1D lattices are both three-dimensional, allowing the electrons to diffract in directions not restricted to the dimensionality of the real lattice. The reciprocal lattice and Brillouin zone construction for a 2D real lattice are shown in Fig. 5.5. Since the reciprocal lattice consists of rods perpendicular to the sample surface, the reciprocal-lattice vectors g form a continuum. So do their bisecting planes. If the momentum transfer is along the (00) rod, as in Fig. 5.5a, the Brillouin zone degenerates to a continuum, and no Kikuchi lines can be seen. The case is different, however, if the momentum transfer of the diffraction process involves different reciprocal-lattice rods, as shown in Fig. 5.5b. Now the Brillouin zone boundaries no longer cover all space, but form a parabolic envelope. This construction corresponds to the definition of the parabola as a curve with equal distances from a point and a line, with the reciprocal-lattice rod defining the directrix of the parabola. As in the 3D case, these parabolic surfaces extend to infinity in the directions perpendicular to g_{hk}. The parabolic envelopes of the Brillouin zone boundaries are more intense than the area outside, so that they become visible as lines when we apply the same construction procedure for Kikuchi lines as in the 3D case. When we shift k' and k_0 by $-g$ and interchange k' with k_0, we obtain the same construction, namely that the intersection of a sphere of radius k_0 around (000) with the Brillouin zone boundaries defines the Kikuchi line pattern. In the 2D case, the pattern consists of parabolas opening parallel

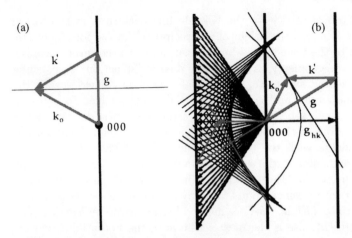

Fig. 5.5. Brillouin zone construction and Kikuchi lines for a two-dimensional lattice. The case with momentum transfer g along a rod is given in (**a**). Part (**b**) shows the case for g involving two different rods differing by a two-dimensional reciprocal-lattice vector g_{hk}

to the shadow edge with (000) as their common focus. Again the translation properties and the inversion symmetry with respect to (000) ensure that the diffracted vectors k' define the same pattern as the construction using the sphere of reflections with k_0.

True parabolic Kikuchi lines are difficult to find in experimental patterns, since real lattices are always three-dimensional, even if one dimension of the sampled volume is very small, as in RHEED. In addition, the 3D reciprocal-lattice points all lie on the 2D streaks, so that they form part of the parabolic envelope. Therefore, lines corresponding to maxima on the streaks are enhanced and the envelope is not continuous. A good approximation to the two-dimensional case is shown in Fig. 5.6, where a surface cleaved in situ in the MBE chamber was used to obtain a microscopically smooth surface. The parabolas simulated for a potential of 6 V constitute a good approximation to the observed pattern. All Kikuchi lines run close to the parabolic envelopes. The agreement is worst for the first-order lines, where the pattern shows an asymmetry, probably due to diagonal cleavage steps on the surface. For comparison, the corresponding 3D simulation is shown in Fig. 5.7. It is evident that line segments far from the parabolic envelopes are absent in the experimental pattern. In this sample, the Kikuchi line formation process is therefore closer to the description using a two-dimensional lattice.

The construction scheme for Kikuchi lines can be extended to 1D crystals by recognizing that the reciprocal lattice for one-dimensionally confined electrons consists of planes perpendicular to the confined electron beam. The corresponding geometry is contained in Fig. 5.5b, if we identify the electron direction with the vector g_{hk} running horizontally in the plane of the figure

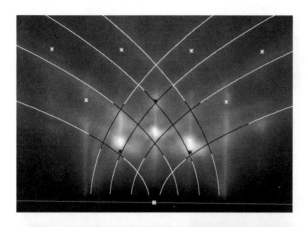

Fig. 5.6. RHEED pattern of a GaAs (110) surface cleaved in UHV along the [110] direction. The *lines* indicate a 2D Kikuchi line simulation for a potential of 6 V

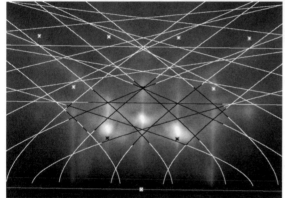

Fig. 5.7. The same experimental pattern as in Fig. 5.6, but with a 3D simulation with a potential of 10 V

and consider the vertical lines as planes perpendicular to the image plane. The Brillouin zone envelopes form confocal paraboloids of revolution and the construction using the sphere of reflection produces circular Kikuchi lines. These lines are not observed in RHEED, since there are no Brillouin zone boundaries running parallel to the confinement direction. Nevertheless, 'circular' Kikuchi lines due to one-dimensional channeling effects can be observed in favorable cases. An example is shown in Fig. 5.8, taken from a (112)A surface close to the [1$\bar{1}\bar{1}$] direction. In this case, the circular line segments are not due to paraboloids of revolution, but constitute the envelope of several parabolas similarly to the transition from the 3D to the 2D case. The crystal exhibits a sixfold rotational symmetry along the [1$\bar{1}\bar{1}$] axis. An electron traveling along this axis can be thought of being confined in three planes simultaneously. The construction of the circular line therefore corresponds to rotating the parabolic pattern of Fig. 5.5b twice by 60° around the zone axis of the crystal and superimposing the three patterns. The envelope of the six parabolas is a good approximation to a circle and forms the pattern shown in Fig. 5.8.

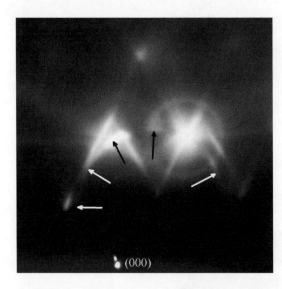

Fig. 5.8. Example of a circular Kikuchi line due to one-dimensional channeling. The pattern was recorded from a GaAs (112) surface close to the $[1\bar{1}\bar{1}]$ zone axis

5.2 Determination of Average Crystal Potential and Misorientation

The simulation of Kikuchi lines provides us with an independent method to determine the average crystal potential as well as the surface misorientation. The assumption of randomized beam directions after the first scattering event decouples the Kikuchi pattern from all coherent diffraction processes. We therefore get two different, simultaneous diffraction measurements with every RHEED pattern. This provides us with additional, independent information that can be used to cross-check results. We can, for example, confirm the surface misorientation deduced from the splitting of the elastic peak by comparing it to the misorientation angle obtained from a Kikuchi line fit.

In practice, Kikuchi processes are often difficult to distinguish from elastic scattering. This is especially difficult when the Ewald sphere tangentially touches a rod so that the diffracted vector lies in the surface plane. In these cases of surface resonance some of the Kikuchi scattering paths are superimposed on the elastic scattering pattern. The intensities of the elastic pattern can then be strongly altered, since the diffracted Kikuchi wavevectors again end on the Ewald sphere. For our purposes, however, these special cases do not matter, since we only use the positions of certain features, not their intensities, for our analysis. Within the measurement accuracy, the geometry remains the same whether the first scattering event of the Kikuchi process is elastic or inelastic with an energy loss in the sub-eV or even several eV range.

The direct measurement of the average crystal potential allows us to check the results of dynamical calculations, where the average crystal potential is generally treated as a fitting parameter [73]. Two examples taken from a

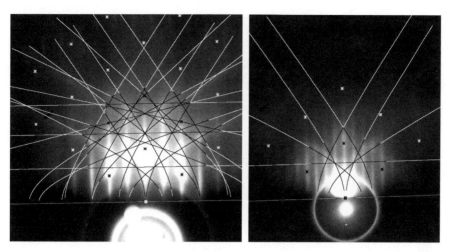

Fig. 5.9. Kikuchi line simulations compared with an AlAs (001) $c(4\times4)$ pattern (*left*) and a GaAs (001) $\beta(2\times4)$ pattern at a very low incidence angle (*right*). Both images were taken along [$\bar{1}10$] with 20 keV electrons

large number of fits are shown in Fig. 5.9. From the AlAs surface in the left-hand panel of Fig. 5.9, the potential was determined as 10.4 V; the fit in the right-hand panel for GaAs $\beta(2\times4)$ gave a value of 10.7 V. Generally, potential values of 10.5 V $\pm$ 1 V are found for both GaAs and AlAs surfaces, for a variety of surface orientations and observation azimuths. These values are significantly lower than those used in the theoretical calculations, which are between 13 and 14 V [73]. Possible reasons for this discrepancy will be discussed in Sect. 5.3. The surface reconstruction is found not to affect the measured potential values within the accuracy achieved.

The Kikuchi line fit also allows a reliable determination of the sample miscut. An example of a GaAs (001) surface misoriented predominantly towards [$\bar{1}10$] is shown in Fig. 5.10. The surface is reconstructed into the higher-temperature $\alpha(2\times4)$ structure with very weak half-order streaks. In this regime, the Kikuchi lines show more contrast, allowing a good adaptation of the calculated pattern. Both the miscut and the potential were fitted in an alternating iterative sequence. The miscut determined from the left-hand panel of Fig. 5.10 is 1.62°, with an average potential of 10.9 V. The fit in the right-hand panel yields −1.86° with 10.5 V. The data obtained from the right-hand panel are less reliable because of the weaker contrast of the pattern.

From fits to similar surfaces, we find generally that the angles determined in the 'uphill' geometry, corresponding to the right-hand panel of Fig. 5.10, are larger than the 'downhill' values. This degrades the accuracy of the method. Putting more weight on the 'downhill' azimuth, we therefore estimate the accuracy of the method as ±0.1°. Note that a slight deviation of the

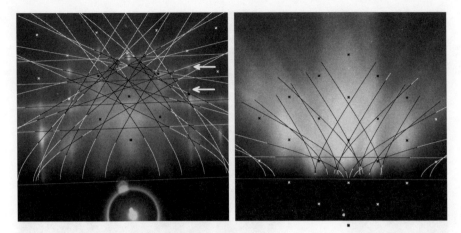

Fig. 5.10. Miscut GaAs (001) surface along [$\bar{1}10$] and [$1\bar{1}0$]. The misorientation angle can be determined from the Kikuchi line fit by means of the distance between the reciprocal-lattice origin and the shadow edge

azimuthal angle does not hamper the fit, as demonstrated in the right-hand panel of Fig. 5.10. As long as angular distortions are small, the lateral spacing of the rods necessary for the determination of the overall scaling factor is still accurate, while the perpendicular distances of characteristic features from the shadow edge are independent of rotation for small angles.

A closer look at the left-hand panel reveals that the Kikuchi pattern is slightly tilted with respect to the shadow edge. This indicates a lateral miscut of the sample surface in the direction parallel to the screen. For a regularly stepped surface, as treated in Sect. 4.2, the streak is expected to remain perpendicular to the singular planes between the steps when the surface is tilted. However, the streak consists of several subrods that run along the normal to the average miscut surface plane, representing the reciprocal lattice of the step edges [114]. If the Ewald sphere cuts these rods close to an out-of-phase condition, there are either multiple reflections, in the case of a regular step array, or an intensity maximum displaced from the average center of the rod. Owing to the anisotropy of the surface [38], the roughness is larger parallel to the screen than along the beam in the [$\bar{1}10$] direction. In addition, the transfer width is smaller perpendicular to the beam direction, so that the split structure of the rods is not resolved in the direction parallel to the screen in Fig. 5.10. As soon as the split rod structure can be resolved, a miscut-angle determination from the elastic pattern becomes possible. In the angular range below this limit, a determination with the necessary accuracy of the miscut angle parallel to the screen is not possible from the Kikuchi pattern, since the width of the lines in the experimental pattern is generally too large and the shadow edge is usually not very clearly defined.

As an independent check of the results in the direction perpendicular to the screen, the rod splitting for the $[\bar{1}10]$ direction indicated by the two white arrows can be used in a kinematical model to calculate the misorientation angle in the plane of the incident beam [114]. The average surface tilt θ_c is given by

$$\theta_c = \frac{\langle \theta_e \rangle}{2\pi/sd - 1} , \tag{5.2}$$

where $\langle \theta_e \rangle$ represents the average exit angle of the two spots marked by the arrows with respect to the singular crystal plane, d is the planar spacing (2.83 Å for (001) GaAs) and s denotes the spot separation in reciprocal lattice units. For the example of Fig. 5.10, we obtain a value of $1.8 \pm 0.1°$, in good agreement with the Kikuchi line method.

Another interesting phenomenon occurs when we investigate the Kikuchi line pattern of the (113)A surface. A simulation for a potential of 10.2 V and a misorientation angle of 0.78° is shown in Fig. 5.11. The horizontal Kikuchi line is split as indicated by the arrows in Fig. 4.8. Whereas the lower line of the doublet agrees with the other lines of the simulated pattern, the upper line corresponds to a simulation with zero potential. This phenomenon has been observed previously in transmission reflection high energy electron diffraction (TRHEED) [225], where the incident beam is directed to a position close to the sample edge so that some of the diffracted beams exit at the side face of the sample crystal.

The mechanism responsible for the splitting can be explained by the exit geometry of the Kikuchi electrons at a vicinal surface   as shown in

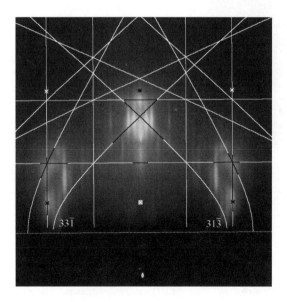

Fig. 5.11. Simulated Kikuchi line pattern for the misoriented GaAs (113)A surface of Fig. 4.8 using a potential of 10.2 V. The simulation is adapted to the lower line of the Kikuchi line doublet of Fig. 4.8 and the other lines agree well

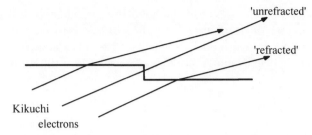

'unrefracted'

'refracted'

Kikuchi

electrons

Fig. 5.12. Ray paths of electrons leaving the crystal through terrace planes and step edges. The difference in refraction leads to split lines

Fig. 5.12 [222]. When the electron crosses the potential step, its momentum perpendicular to the local surface orientation is reduced by the amount given in (5.1). If the electron exits at a step face, the momentum change is along the beam, and its direction remains basically unchanged, whereas a significant deviation occurs for an exit path through the terrace top face. The two lines of the doublet therefore relate to different scattering geometries, and comparable intensities of the 'unrefracted' and 'refracted' lines indicate a large number of facets perpendicular to the beam direction on the surface.

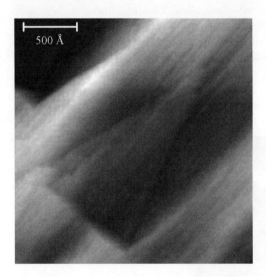

500 Å

Fig. 5.13. GaAs (113)A surface misoriented towards [$\bar{3}\bar{3}2$] by 0.9° like the sample used in Fig. 5.11. STM image taken by M. Wassermeier

The presence of these facets on the misoriented surface can be verified by STM as shown in Fig. 5.13. The surface exhibits a characteristic stepped structure with triangularly shaped features that accommodate the miscut in multilayer steps perpendicular to the [$\bar{3}\bar{3}2$] direction. The 'downhill' direction points towards the upper right corner of the picture. The $\langle 212 \rangle$ side edges of the triangles follow the angle of the primitive unit cell shown in Fig. 1.6c. From the STM image, it is not possible to determine whether the faces of the steps perpendicular to [$\bar{3}\bar{3}2$] are {111} or {110} planes, because the radius of the tip [226] does not permit access to these regions. It is not yet clear whether the triangular features are intrinsic to vicinal growth on this surface or whether they are connected to structural defects generated at the substrate–film interface. Since the size of the features grows with increasing film thickness, the latter seems more likely. The surface shows significant roughness perpendicular to the $[\bar{3}\bar{3}2]$ direction. At the same time, outside the triangular areas corresponding to singular (113) terraces, there are regions where the surface is regularly stepped. Therefore, the STM image does not allow a distinction between the two mechanisms that can lead to a splitting of the central streak in RHEED as discussed in Sect. 4.2. Most likely, the modulation of the streak is due to a combination of both mechanisms.

5.3 Where do Kikuchi Lines Originate?

So far, we have assumed that the process responsible for the generation of Kikuchi lines takes place away from the surface in the bulk material. A closer look at the simulations, however, leads us to question this assumption. For the construction of the patterns in this chapter, it was always sufficient to use only the reciprocal-plane points in the upper half plane above the shadow edge. Lines that correspond to points in the lower half of plane, such as $\bar{1}1\bar{1}$ (see Fig. 5.3 for the indexing scheme), are not observed. Returning to the construction of the sphere of reflections shown in Fig. 5.2, we can arbitrarily divide it into two half spheres along any plane that contains the origin. If we then use the scheme given there, we note that for the construction of the lines in one half sphere, only reciprocal-lattice points located in the other half sphere are needed, because of the translation step. The absence of lines belonging to reflections in the other half sphere is therefore a direct consequence of the truncation of the crystal. The reciprocal lattice needed for the construction of the Kikuchi line pattern shows the same truncation as the real lattice it belongs to, which means that the reciprocal-lattice points in one half sphere correspond to real-space points in the same half sphere, at least for cubic crystal symmetry. We can therefore identify Kikuchi line generation in RHEED as a surface-related process.

Owing to the similarity of the atomic form factors of Ga and As, the reciprocal lattice of GaAs approximates the diamond reciprocal lattice. This is evident from the weakness of reflections such as (002) that are forbidden in the diamond structure. We therefore expect Kikuchi lines that correspond to these only weakly allowed reflections to be practically invisible in the RHEED pattern. A simulation assuming diamond symmetry of the lattice is shown in Fig. 5.14. One can clearly see that not all lines visible in the pattern are reproduced by the simulation. These visible lines are the 006 and, especially, the $2\bar{2}6$ and $\bar{2}26$ reflections, which are as strong as the allowed reflections. This provides further evidence that Kikuchi lines are generated in the surface layer, where the extinction of certain reflections is lifted. This should lead to a variation of the relative intensities of Kikuchi lines as a function of the

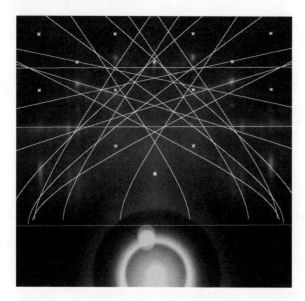

Fig. 5.14. Simulated Kikuchi line pattern for a miscut (001) GaAs (2×4) surface assuming diamond symmetry of the lattice, which should be a good approximation for a non-truncated GaAs crystal

surface reconstruction. Comparison of the two parts of Fig. 5.9 suggests that such an effect is present; this was not, however, investigated in detail.

The values of the potential obtained from our fitting procedure are consistently between 10 and 11 V, which are significantly lower than the values of 13 to 14 V generally used in dynamical RHEED theory for GaAs [73]. In the present calculations, the crystal potential obtained from the Fourier transform of the Doyle–Turner potentials [108] was adjusted by an additive offset of about 1 V as a fitting parameter in the simulations. Since our method provides an accuracy of about ±1 V, the discrepancy cannot be explained by experimental uncertainties. The theoretical values are based on the Doyle–Turner potentials, which, as was pointed out in Sect. 4.3, overemphasize the forward scattering direction compared with experiment in the kinematical simulations. On the other hand, these potentials are widely used with high accuracy, for example in pattern modeling for TEM. If we therefore accept the 13–14 V value, a measured potential of 10–11 V again means that the Kikuchi pattern originates very close to the surface. If we assume the 13–14 V value for the bulk structure, this means that the Kikuchi diffraction process takes place in the reconstructed layers or close to steps and vacancies that reduce the average potential. For the top reconstructed layer, in the kinematical simulations of Sect. 4.3, we obtained a value of about 7 V. A value of the potential of 10 V could therefore be roughly assigned to a position just below the topmost surface layer.

In the construction of the Kikuchi line pattern, we assume kinematical scattering in the second step of the process. With the large cross-sections involved in electron diffraction [104], however, kinematical scattering theory is only valid for relatively short distances. The extinction distance, which can be taken as an upper limit after which kinematical reflections vanish and multiple scattering dominates, can be estimated as roughly 1500 Å for electron energies of 20 kV [227]. For an exit angle of 2°, this corresponds to 9 ML below the surface. From the semikinematical simulations of Sect. 4.3, we found that third-layer atoms already do not contribute to the elastic pattern. We interpreted this as a dominance of dynamical scattering for scatterers below the second layer. These extinction arguments therefore provide an upper depth limit of several ML below the surface for Kikuchi line generation.

Even for surfaces that are rougher than the ones in Fig. 5.6, high-intensity lines are generally observed only close to the 2D parabolic envelopes. This indicates that the primary scattering event in the Kikuchi process leads to a preferred distribution of the electrons in directions closely parallel to the surface instead of an isotropic randomization of their directions. The Kikuchi lines observed in RHEED, therefore, are 2D rather than 3D in nature. Owing to the construction of the reciprocal-lattice vectors, every Kikuchi line corresponds to a real-space lattice plane. We can therefore speculate that the selection of lines that is observed in the RHEED pattern corresponds to the set of microscopic surface facets or steps that are actually realized on the real-space surface. This could explain the absence of line splitting for the diagonal $31\bar{3}$ and $33\bar{1}$ lines in Fig. 5.11. The surface regions terminating in the large multilayer steps perpendicular to $[\bar{3}\bar{3}2]$ are relatively flat. Therefore, the steep {331} planes do not terminate at these steps to produce 'unrefracted' lines of this orientation. Large-scale scans of the same surface show regions in between the triangles with steep lateral inclination that can contain the {331} orientations of the diffracted lines. Exactly oriented GaAs (113) samples exhibit a much smoother morphology. On these surfaces, the Kikuchi lines belonging to the {331} planes are much weaker, as can be readily verified by inspection of Fig. 4.1.

We conclude that several experimental facts point towards a surface-sensitive model for the generation of Kikuchi lines in the RHEED geometry. In this model, the construction scheme is similar to the bulk diffraction model with the difference that only the planes that are actually realized as part of the surface morphology take part in the pattern generation process. Such a model can explain the characteristic absence of line segments away from the parabolic envelopes as well as the presence of quasi-forbidden lines. It is also in agreement with the observed small value of the $10.5 \, \text{V} \pm 1.0 \, \text{V}$ potential correction for the exiting electrons.

6. RHEED with Rotating Substrates

Except for the long-time exposures in Figs. 4.14 and 4.18, the RHEED measurements described in this book up to this point were performed without moving the specimen. This is still the standard way of doing RHEED experiments. In MBE, however, the sample is rotated around a surface-normal axis during crystal growth for uniformity reasons. Despite the fact that RHEED has been used in MBE for quite some time, its main applications, namely growth rate calibration and surface structure determination, are still performed with the substrate held at a fixed position. For the growth of device structures, the substrate is then rotated and the RHEED beam is turned off or the continuously changing diffraction pattern during rotation is used merely for a rough, qualitative visual inspection.

There have been several attempts to resolve this incompatibility. One way of measuring RHEED on a rotating sample is to measure the specular spot intensity as a function of azimuthal angle. The resulting azimuthal plots [228–231] can be fitted with high accuracy using dynamical theory. This works best if the data are acquired at generalized Bragg conditions (i.e. taking into account refraction effects, see Sects. 4.3.2 and 9.2.1). The wobble and misorientation of the substrate have to be negligibly low for one to get meaningful results. The good fit between experiment and calculations allows stringent tests of proposed surface structures if high-quality experimental data can be obtained. Most MBE sample manipulators, however, do not allow sufficiently precise movement of the sample to apply this measurement method as a standard technique during growth.

Monitoring the specular spot intensity, as well as its full width at half maximum (FWHM), during a complete growth sequence allows a correlation between the signal obtained and the layer structure [232]. The signal is low-pass filtered to eliminate the frequencies associated with the substrate rotation. This also eliminates the growth oscillations. Since the correlation with the azimuthal angle is lost as well, no analysis in terms of diffraction theory is possible. The RHEED signal, however, can be successfully correlated with device performance from run to run.

In the following sections, we will develop and discuss three additional, promising approaches, both to access reciprocal space and to measure RHEED oscillations on rotating substrates. All of them rely on phase-locked substrate rotation combined with a high data acquisition rate to obtain useful results. To obtain a phase lock between substrate rotation and diffraction intensity detection, the substrate rotation motor of the MBE system is interlocked with the timebase signal of the CCD camera acquiring the diffraction pattern. The ratio of motor revolutions to camera frames can then be chosen so that one substrate rotation corresponds to an integer number of acquired data points. By running a standard TV-frequency CCD camera in noninterlaced field mode and modifying the image digitizer accordingly, data can be acquired at a rate of 50 Hz. This turns out to be sufficient for most measurements and avoids the use of expensive, nonstandard high-speed CCD cameras.

6.1 Gated Detection

A straightforward way to obtain information similar to a static experiment is the acquisition of still frames at fixed azimuths during the continuous rotation of the substrate. Examples of the resulting RHEED patterns are shown in Fig. 6.1. In this measurement, the data were recorded at angular spacings of $90°$. The four main azimuths of a GaAs (001) $\beta(2\times4)$ surface are shown. The left-hand column shows the first revolution, the central column was recorded

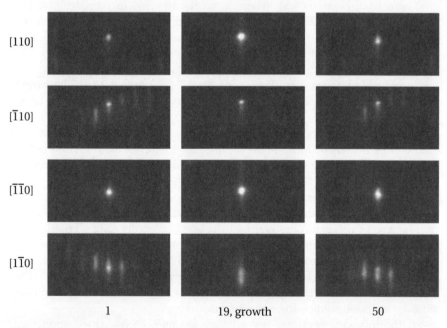

Fig. 6.1. Synchronized snapshots of the four main azimuths of $\beta(2\times4)$ reconstructed GaAs. The first, 19th and 50th rotations while rotating the sample at 0.25 Hz are shown. The *central column* was measured during GaAs homoepitaxy

in the 19th revolution during GaAs homoepitaxy, and the right-hand column shows the data for the recovered surface in the 50th revolution. This measurement method allows a quasi-simultaneous acquisition along different azimuths, which would correspond to an arrangement with multiple RHEED guns and screens at the respective azimuths [88]. The resulting images can be processed similarly to a static RHEED image, except that the temporal resolution is limited because the images are updated only once per revolution.

Depending on the rotation frequency, the non-(00) streaks become elongated because the corresponding spots move vertically during the exposure time of the CCD. A relatively slow rotation is therefore needed if the Laue circle position needs to be resolved. The measurements of Fig. 6.1 were done in a VG V80 system. The poor dynamic range is due to the low intensity of the standard RHEED gun mounted on this chamber. The sample holder in this machine is kept from turning by a notch. This mechanism allows the sample holder to move slightly back and forth within the sample manipulator during each rotation. This oscillation shows up in the difference between the $[\bar{1}10]$ and $[1\bar{1}0]$ azimuths. The $[\bar{1}10]$ image is slightly to the left and the $[1\bar{1}0]$ is slightly to the right of the exact azimuthal position. This type of gated measurement directly reveals the sample wobble and miscut through the position of the specular spot with respect to the shadow edge, similarly to Sect. 5.2. Although the play in the mechanical gears inside the vacuum chamber and the vacuum feedthrough itself is significant, the azimuthal accuracy of the setup is remarkable. This can be attributed to the cyclical variation of most of these deviations, causing the substrate to rotate in a very similar way in each turn if the data acquisition is phase-locked with the rotation.

6.2 Azimuthal Scans

Gated detection allows access to selected sections through reciprocal space. It would be more desirable, though, to image more complete representations of reciprocal space. Spherical RHEED screens permit the imaging of reciprocal-lattice rods in the vicinity of the (00) rod [93], allowing the determination of the 2D surface lattice from a static RHEED pattern [72]. An almost complete scan of reciprocal space, however, can be achieved by recording continuously during sample rotation [233]. The diffraction geometry for this case is shown in Fig. 6.2. Figure 6.2a shows the standard situation for nonrotating RHEED. The condition of elastic diffraction requires that, for any incidence angle θ, both k_0 and k' end on the (00) reciprocal-lattice rod for diffraction to occur. During sample rotation, the reciprocal lattice revolves around the (00) rod, while the Ewald sphere remains stationary if we neglect wobble. The Ewald sphere therefore scans almost the entire upper half of reciprocal space as indicated by the shaded area in Fig. 6.2b. In principle it is possible to reconstruct almost the entire upper half of reciprocal space from such a measurement, but current computer speeds only allow the processing of a two-dimensional

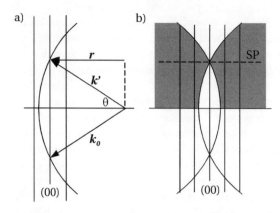

Fig. 6.2. Ewald sphere geometry for (**a**) static RHEED and (**b**) during rotation in the coordinate system of the rotating crystal. The *shaded toroidal region* in (**b**) with rotational symmetry around (00) represents the reciprocal-space volume which is accessible during sample rotation. The plane parallel to the surface and intersecting the specular spot, the specular plane, is indicated by a *dashed line*

data set in near-realtime. The planes perpendicular to the sample surface are approximated reasonably well by the gated-detection slices of the previous section. We therefore now focus on the planes parallel to the surface that directly reveal the full in-plane symmetry of the surface. Such scans are impossible with nonrotating RHEED.

The only surface-parallel plane within the donut-shaped accessible volume that does not have a void in the center is the plane through the specular spot. We call this plane the specular plane (SP). Its position is marked in Fig. 6.2b by a dashed line. The measurement is performed by measuring the intensities along a line parallel to the sample surface through the specular spot, as shown in Fig. 6.3a. The specular plane is then reconstructed by plotting these lines, bent with radius r (see Fig. 6.2a) and centered on the specu-

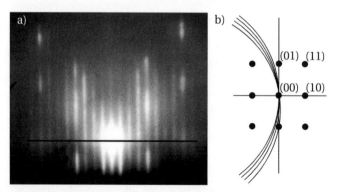

Fig. 6.3. Measurement geometry and top view of the azimuthal scan. The intensities along the *line* indicated in (**a**) are measured as a function of azimuthal angle. The intensity distribution within the specular plane is then obtained by plotting the lines at their respective angular positions with radius r (see Fig. 6.2) and centered at (00) (see text)

lar spot, according to their azimuthal angle. This is schematically shown in Fig. 6.3b. As the rotation proceeds, the lines trace the entire plane so as to represent a section through reciprocal space at a distance $2k_0 \cos \theta$ from the origin of the reciprocal lattice defined by the tip of $\boldsymbol{k}_0$. Such a section is very similar to a LEED pattern. By varying the incidence angle, a wide range of surface-normal reciprocal-lattice vectors can be probed, similarly to an I–V measurement in LEED. The raw data of the linescan before processing are shown in Fig. 6.4. Both the angles and the corresponding symmetry along

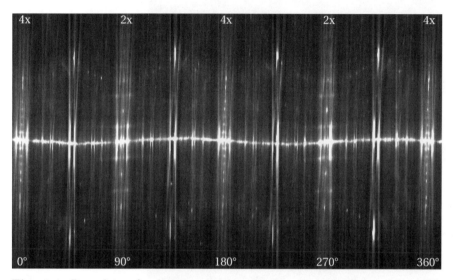

Fig. 6.4. Raw data of a linescan in the specular plane during rotation. Approximately one rotation of a GaAs $\beta(2\times4)$ surface is shown. The reflections on reciprocal-lattice rows not containing (00) lie on curves $y = a \sec bx$

the azimuth are indicated. These data were collected using a low-wobble substrate manipulator. The movement of the specular spot is therefore minimized, though still clearly visible. As long as distances and angles in the SP do not need to be analyzed quantitatively, quite large values of wobble and miscut can be tolerated without destroying the symmetry of the resulting reciprocal space map.

As a demonstration, the SPs of GaAs (001) $\beta(2\times4)$ and $c(4\times4)$ are shown in Figs. 6.5 and 6.6. The rotation speed was approximately 3 rpm. Together with the acquisition frequency of 50 Hz, this results in an angular resolution of approximately $0.4°$. The incidence angle in both cases was $1.5°$, using 20 keV electrons. In both figures, the underlying 1×1 mesh of the surface reciprocal lattice is indicated by a white square. The visible area of the SP spans more than six unit cells in each direction. This size is limited only by the electron energy (an increase of the electron energy reduces the re-

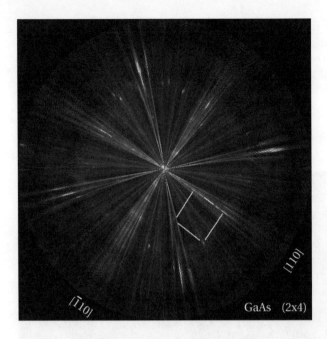

Fig. 6.5. Azimuthal scan of the GaAs (001) $\beta(2\times4)$ surface reconstruction. The surface unit cell and the main azimuths are indicated

ciprocal rod spacing as $1/\sqrt{E}$, see (2.2)–(2.5)) and the size of the RHEED screen. For typical chamber geometries, this amounts to about 16 unit cells

Fig. 6.6. Azimuthal scan of the GaAs (001) $c(4\times4)$ surface reconstruction. The surface unit cell is indicated. All other measurement conditions are identical to those of Fig. 6.5

for GaAs. The eight in-plane $\langle 001 \rangle$ and $\langle 110 \rangle$ azimuths can be easily identified by the adjacent higher-intensity radial streaks. These correspond to perpendicular surface resonance conditions [74,75,234,235] where the Laue circle tangentially touches the reciprocal-lattice rods, leading to an increased total reflectivity.

Bright spots in the azimuthal scan of the SP mark its intersections with reciprocal-lattice rods. The spots are elongated with their symmetry axis in the radial direction. Their anisotropy is a direct measure of the different transfer widths [121] t_w (2.22) in the corresponding directions. Azimuthal scans allow direct access to the high-resolution direction of RHEED parallel to the beam by measuring the SP spot sizes in the tangential direction. In Fig. 6.5, the $\beta(2\times4)$ structure is clearly identified by the rows of closely spaced spots parallel to [110]. This distinguishes it from the $c(4\times4)$ structure in Fig. 6.6, which exhibits a centered mesh. A comparison of the full 2D symmetries of the reciprocal lattices allows an immediate distinction between $c(4\times4)$ and (2×2) structures that is not evident from observations along the $\langle 110 \rangle$ azimuths, in which both structures show the same symmetry.

The $c(4\times4)$ structure in Fig. 6.6 does not have a fourfold rotational symmetry. The spots in the row directly adjacent to the [110] azimuth, for example, show high intensity on the first, third and fifth spots, counting from (00). In the corresponding row along $[1\bar{1}0]$, the third spot is very weak. These deviations from the fourfold symmetry are usually present even on (1×1) reconstructed surfaces and allow the identification of the different $\langle 110 \rangle$ azimuths even if the static RHEED patterns along both directions differ only slightly. This difference is due to anisotropies of the surface, either in the surface morphology or the details of the surface reconstruction such as dimer orientation. Azimuthal scans are therefore promising candidates for comparison with theoretical calculations, since the large number of reflections within a single scan allows a detailed comparison between experimentally measured and theoretically predicted intensities for a given model. Using the kinematical approximation, we found in Sect. 4.1 that the structure of reciprocal space far from the (00) rod contains the the most detailed information about the atomic positions. Given a suitable dynamical model, even the determination of relaxation within a surface reconstruction seems feasible and can be expected to be more reliable than previous optimizations using data sets obtained from RHEED measurements in more conventional geometries [105–107,236].

The comparison of Figs. 6.5 and 6.6 suggests a reliable method for the in-situ control of surface reconstruction. If the growth process requires the surface to maintain a certain reconstruction, the corresponding spots in the azimuthal scan can be monitored during substrate rotation. The two-dimensional nature of the measurement permits the simultaneous monitoring of many spots, resulting in increased accuracy and allowing consistency checks. In this way, a control algorithm could distinguish whether the desired surface reconstruction changes to an adjacent low- or high-temperature phase and, if it controlled the sample temperature, it could change it accordingly.

An example of a surface morphology change is shown in Fig. 6.7. Figure 6.7a shows a (001) InP surface under As$_4$ flux prior to lattice-matched InGaAs deposition; Fig. 6.7b shows the same surface just after growth ini-

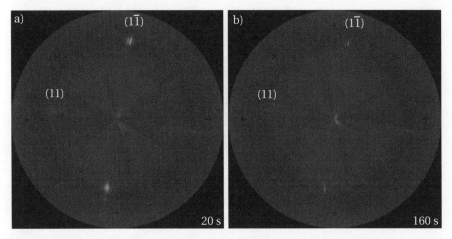

Fig. 6.7. Azimuthal scans of a (001) InP surface before and during InGaAs overgrowth. (**a**) InP surface during the As soak after phosphide growth with the presence of an anisotropic surface structure. In (**b**), at the onset of lattice-matched InGaAs growth, the surface is more isotropic

tiation. The positions of the main reciprocal-lattice rods are indicated. In Fig. 6.7a, the $(1\bar{1})$ and $(\bar{1}1)$ spots show two satellite reflections and are brighter than the (11) and $(\bar{1}\bar{1})$ spots. This is typical for a surface that is rougher along [11] than along $[1\bar{1}]$. The modulation along [11] causes the diffraction spot to develop satellites with the same orientation. These satellites are only resolved in the azimuthal scan; the poor resolution in the radial direction removes the structure on the (11) spot. In Fig. 6.7b, the surface is much more isotropic, the satellites have vanished and all four spots have similar width and brightness.

Phase-locked substrate rotation ensures that the spots in the SP remain stationary. This allows the study of the same reflection in successive rotations. A close-up of the $(1\bar{1})$ spot throughout the growth sequence is shown in Fig. 6.8. The growth analyzed in Figs. 6.7 and 6.8 was performed in the same chamber as was used for the patterns shown in Fig. 6.1, with a standard manipulator. The rotation frequency was 0.05 Hz. The satellite structure represents a section through the tip of the chevron structure frequently seen along $[\bar{1}10]$ in conventional RHEED from rough (001)-oriented surfaces. Note that the central peak is very narrow and well resolved, with the broader satellites clearly separated. When growth is initiated, the satellite peaks van-

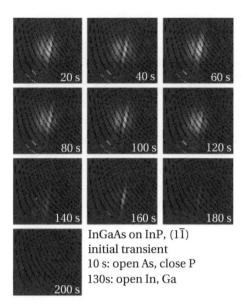

InGaAs on InP, (1$\bar{1}$)
initial transient
10 s: open As, close P
130s: open In, Ga

Fig. 6.8. Time evolution of the (1$\bar{1}$) reflection during the initial stages of InGaAs growth on InP. The phase-locked rotation ensures that the reflection remains stationary in successive azimuthal scans to investigate the peak profile

ish and soon the central peak becomes too weak for further observation. We can therefore deduce that the surface changes from its original strongly anisotropic state through a well-ordered isotropic configuration, after which it becomes disordered again. As long as the RHEED intensities are strong enough, azimuthal scans can be continuously recorded during growth and allow the nonstop characterization of the surface during growth.

6.3 RHEED Oscillations

Although the measurement of growth oscillations is still the main application of RHEED in MBE, the number of growth oscillation studies on rotating substrates is limited. This is mainly due to the high noise levels introduced by mechanical vibrations and the periodic modulation of the signal with the rotation frequency and its harmonics [160]. Both of these mechanisms contribute strongly to the frequency range of the growth oscillations and complicate the reliable determination of the growth frequency. Whereas azimuthal scans allow access to large volumes of reciprocal space, their temporal resolution is limited by the rotation frequency. Azimuthal scans are therefore too slow to measure growth oscillations with typical frequencies around 1 Hz. The only part of the diffraction pattern that is visible continuously during rotation is the (00) rod. The specular spot moves up and down along this rod, depending on substrate miscut and manipulator wobble. Its intensity typically varies by about three orders of magnitude during one rotation for a nongrowing surface [228]. In addition, both the phase and the amplitude of RHEED intensity oscillations depend strongly on the azimuthal [168,237] and polar [238,239] (see Sect. 9.2.2) angles of the electron beam. This leads to ill-defined growth frequencies and beating in intensity oscillation measurements during rotation.

On the other hand, substrate rotation has the advantage of averaging the growth rate across the wafer, thereby leading to reduced damping of the RHEED oscillations (see Sect. 3.1). This has the additional advantage that growth rate calibrations with rotation are much less position-dependent. On a nonrotating sample, only the growth rate at the rotation axis is equal to the growth rate during rotation. The individual effusion cells can have significant flux gradients across the sample that are averaged during sample rotation. Relatively small deviations from the center position can therefore lead to significant discrepancies between the growth rates determined without rotation and the actual averaged growth rate during rotated growth.

In one approach [161], the substrate is therefore rotated at high speeds, resulting in an averaging of these variations on a timescale much smaller than the growth oscillation period. This method works well from a diffraction point of view, but leads to excessive wear of the rotation drive and has not been widely adopted. A different method uses a spot-tracking algorithm combined with numerical filtering techniques to determine the specular spot size and measure its variations during growth [240]. The signal in this case is quite noisy and a narrow filter window needs to be applied to determine the growth frequency, which requires that the growth frequency has to be known quite accurately in advance. The use of the specular spot size or FWHM, though, has several advantages. It typically varies by not more than one order of magnitude during the measurement, allowing simple and fast processing of the digitized data. Also, the specular spot size does not vary as much as its intensity with changing diffraction conditions for the same surface.

Fig. 6.9. The measurement of RHEED oscillations on rotating substrates. (**a**) Position of the measurement line with respect to the RHEED pattern. The peak profile measured as a function of time is shown in (**b**). Time is measured along the horizontal axis. The growth interval is marked by the black bar. One period of the specular-spot trace corresponds to one substrate rotation

Here, we extend this approach by making use of the strongly anisotropic sensitivity of RHEED parallel and perpendicular to the beam along the surface. Instead of measuring the total spot size, we study its FWHM perpendicular to the surface along the (00) beam. The measurement geometry is shown in Fig. 6.9. Owing to the small angle the Ewald sphere makes with the (00) rod, the resulting signal is already amplified by a factor of approximately $\cot \theta$, where θ is the angle the electron beam makes with the sample surface. This leads to a much better signal-to-noise ratio. The FWHM determination algorithm along the line compensates for substrate wobble and misorientation without having to implement a 2D spot-tracking algorithm. The experiments were again performed in a V80 chamber with a standard manipulator on 2 inch GaAs (001) substrates with a data acquisition frequency of 50 Hz. A measurement with a relatively low rotation frequency is shown in Fig. 6.10. The substrate misorientation and wobble are small enough that the specular

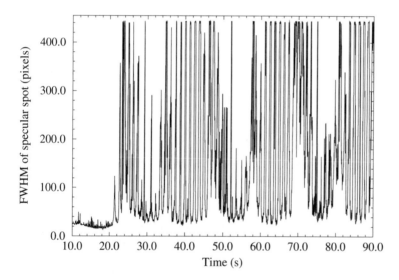

Fig. 6.10. FWHM of the specular spot perpendicular to the substrate as a function of time, in units of CCD pixels. Growth starts at 20 s. The rotation frequency is approximately 0.05 Hz. Owing to small wobble and misorientation, the noise level is low enough to allow one to directly count the oscillations

spot is visible continuously during the rotation. This results in clear data that allow the continuous counting of individual oscillations. The FWHM algorithm is designed to obtain subpixel accuracy by performing a linear fit for the half-maximum points on the shoulders of the peak. It defaults to the length of the measurement line if a reliable FWHM determination is not possible. This leads to a good modulation of the signal since the largest FWHM is usually correlated with low peak intensities.

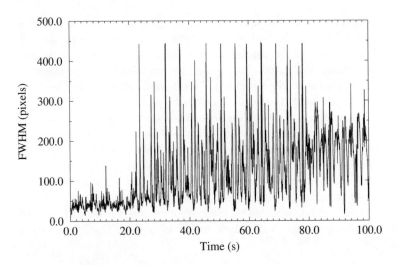

Fig. 6.11. FWHM of the specular spot perpendicular to the substrate as a function of time, in units of CCD pixels. Growth starts at 20 s and terminates at 80 s. The rotation frequency is 0.25 Hz; the growth frequency is 0.80 Hz

With increasing rotation frequency, the signal becomes more noisy, as shown in Fig. 6.11. Within the finite exposure time of the single video frame, the diffraction pattern now changes appreciably, leading to peak broadening during the rotation segments in which the specular spot moves faster. Also, the amplitude of the mechanical vibrations of the substrate increases. Additional noise is introduced if the specular spot is not continuously visible during a complete rotation. In these cases, the direct counting of the oscillations is less practical and a Fourier transformation of the signal leads to a more accurate measurement. The Fourier transforms (magnitudes) of three different growths are superimposed in Fig. 6.12. The three measurements differ only in the GaAs growth rates, indicated by the Ga cell temperatures; all other parameters are identical. The rotation was phase-locked to the RHEED measurement. The spectrum consists of the rotation frequency (0.16 Hz) and its harmonics as well as the growth-frequency peaks around 0.8 Hz. The growth-frequency peaks are accompanied by satellite peaks spaced at multiples of the rotation frequency. The accuracy of the growth frequency as determined by the FWHM of the frequency peak is usually 1 % for measurement times above 150 s and growth frequencies around 1 Hz. Since the rotation frequency has an *exact* value due to the phase-locked rotation, the positions of the satellite peaks can be determined in addition to the central peak. Usually, at least two satellites on either side can be distinguished. In the measurement of Fig. 6.12, even the fourth-order satellites on the low-frequency side are visible around 0.12 Hz. The inclusion of these satellites increases the measurement accuracy as $\sqrt{n}$, since it allows the averaging of all n peak positions. For the mea-

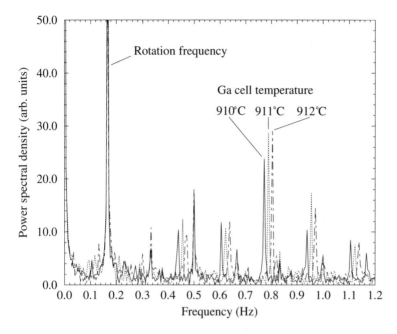

Fig. 6.12. Frequency spectra of FWHM measurements at three different GaAs growth rates, but an identical rotation frequency of 0.16 Hz. The measurement time was 180 s. The growth frequency and its satellites can be clearly distinguished by their displacement with changing Ga cell temperature

surement shown, this results in a total accuracy of less than 0.5 % or $\pm0.1°$, which is about the standard accuracy of MBE effusion cell heaters.

Why is this method better than the traditional measurement of intensity oscillations? In the following section, we compare the intensity measurement of the specular spot area on a nonrotating substrate with the intensity signal integrated over the line along the (00) rod and the FWHM measurement along this line. The three signals are shown in Fig. 6.13 for identical growth conditions of GaAs (001) homoepitaxy and a growth time of 160 s. The top line represents the traditional measurement of the integrated spot intensity during growth without substrate rotation. The center and bottom curves were processed from the same linescan along the (00) streak. In the center curve, the intensity along the line was integrated, whereas the bottom curve shows the FWHM along the line similarly to Figs. 6.10 and 6.11. Compared to the signals obtained during rotation, the top trace shows much stronger damping, due to variations of the growth rate along the probed surface area [158]. This type of damping is strongly reduced during rotation. For the integrated intensity during rotation, however, the noise level also increases strongly, resulting in about the same number of usable oscillations until the signal vanishes in the noise. The FWHM signal, on the other hand, shows a good

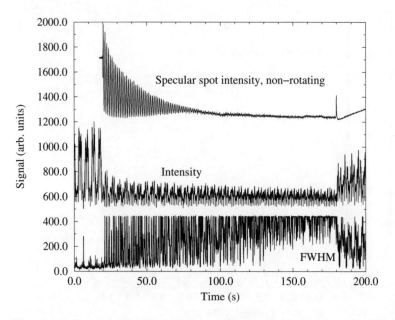

Fig. 6.13. Direct comparison of the oscillation signals obtained by three different measurement methods. The *top trace* shows the integrated specular spot intensity from a nonrotating substrate, the *middle* and *bottom curves* show the intensity integrated along the measurement line and the FWHM of the specular spot, respectively. The bottom two curves were obtained from the same raw data

signal-to-noise ratio, and the crossover between signal and noise is reached after typically twice the number of oscillations compared to the nonrotating intensity signal.

Note that in addition to the strong damping, the average signal of the top trace is not constant, but drops significantly as growth proceeds until a steady state is reached. The fit of a sinusoidal function, therefore, would have to include additional fitting parameters besides the frequency and damping, making it difficult to perfom reliably in realtime. In comparison, the FWHM signal has a constant amplitude without damping and drift, at least out to about 100 s in Fig. 6.13. In addition, the amplitude is predetermined from the length of the measurement line and is therefore fixed. A fitting of the curve should therefore be much easier and more reliable than for the intensity data, since only the frequency and phase remain as parameters..

Another important factor in the measurements on rotating substrates is the suppression of the rotation-frequency components in the signal. Figure 6.14 shows an enlarged section of Fig. 6.13 at the beginning of growth. The dashed lines indicate the period of the growth oscillations. The fourfold rotational symmetry of the cubic crystal structure underlying the GaAs (001) surface modulates any signal measured during rotation with a frequency of four times the rotation frequency. The rotating intensity signal (center curve)

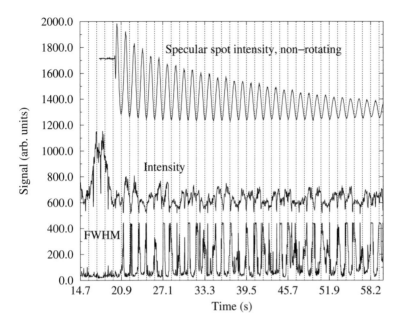

Fig. 6.14. Enlarged section of Fig. 6.13. The modulation and damping of the oscillation signal are significantly different in the three cases

therefore coincides with the growth frequency roughly every 11 growth oscillations. This oscillation, with a slightly higher frequency, is also present in the FWHM signal, but in contrast to the intensity data there is also a strong peak at each marker line. In the intensity measurement, minima at the dashed-line positions are weak if they do not coincide with the rotation minima. This is not the case in the FWHM measurement. The suppression of the rotation frequency and its harmonics is therefore better in the FWHM measurement.

This difference immediately becomes apparent in frequency space. The Fourier transforms of the measurements of Fig. 6.13 are shown in Fig. 6.15. The solid line represents the nonrotating intensity signal, the dashed line is from the rotating intensity signal and the dotted line is the transform of the rotating FWHM signal. All traces are normalized to the magnitude of the growth oscillation peak. The fourfold-rotation-frequency peak is ten times stronger for the intensity measurement during rotation than for the FWHM measurement. In addition to the strong rotation-frequency suppression, the FWHM measurement also has the best frequency resolution, as determined by the FWHM of the growth oscillation peak. The values are 0.027 Hz for the nonrotating intensity, 0.021 Hz for the rotating intensity and 0.013 Hz for the rotating FWHM. The width of this peak is affected by the damping and noise of the time–space signal and the duration of the measurement. Since

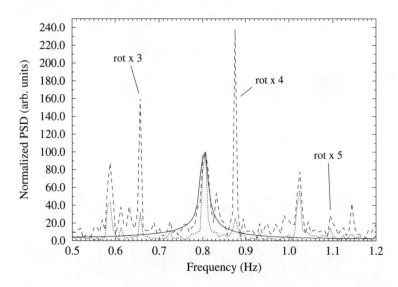

Fig. 6.15. Comparison of the Fourier spectra obtained by transforming the three signals shown in Fig. 6.13. The results are normalized to the growth-frequency peak. The *solid line* corresponds to the nonrotating integrated specular spot intensity, the *dashed line* to the integrated line intensity during rotation and the *dotted line* to the FWHM of the specular spot along the line

the measurement time for all three curves was identical, the broadening of the nonrotating intensity peak is due to the strong damping of the signal [159]. On the basis of the present data, we cannot decide whether the difference between the two rotating-signal peaks is due to the different noise levels or due to the differences in damping.

To study the dependence of the FWHM measurement on the measurement direction across the specular spot, FWHM measurements of GaAs homoepitaxy were performed parallel and perpendicular to the surface on a nonrotating substrate. The time-domain results are shown in Fig. 6.16. The geometry and the frequency-domain data are shown in Fig. 6.17. Since the Ga shutter was magnetically coupled, the specular spot moved to the left during growth and was centered on line 7 in the inset of Fig. 6.17, resulting in large noise levels before and after growth in trace 7. The raw data, plotted on the same scale in Fig. 6.16, clearly show the intrinsic amplification of the signal measured perpendicular to the substrate surface. In addition, the signals parallel to the surface are more strongly damped, combined with an initial increase for curves 1 and 2. These amplitude variations should broaden the frequency peaks. The short measurement time of 40 s, however, results in uncertainty-relation limited peak widths in Fig. 6.17 and, with the exception of line 6, these differences cannot be resolved. It is clear, however, that the absolute noise levels for the different measurement directions are comparable

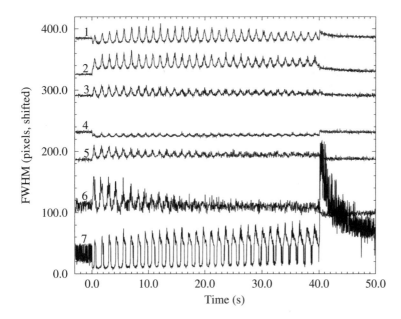

Fig. 6.16. FWHM signals taken perpendicular and parallel to the substrate surface at different positions along the streak. The substrate was not rotating. The positions of the measurement lines are shown in the inset of Fig. 6.17

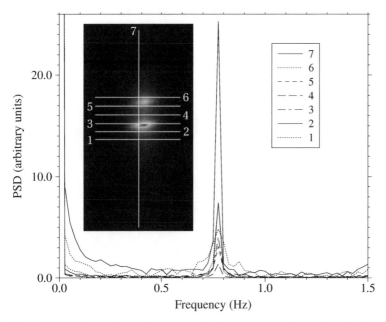

Fig. 6.17. Comparison of the growth-frequency peak magnitudes for the different FWHM signals measured along the lines shown in the inset. Clearly, the measurement perpendicular to the surface (7) results in the largest peak

and therefore the signal-to-noise ratio of curve 7 is better because of its larger absolute signal, despite the fact that the spot is split in this direction.

Theoretical calculations for the (01) and (00) spot sizes perpendicular to the surface as a function of step density [145–147, 241] indicate an increase in the FWHM as a function of step density. This suggests that the FWHM of the specular spot is more directly related to the surface step density than its intensity (see Sect. 9.2.2), which may explain the better accuracy of the FWHM experiments by their reduced sensitivity to changing diffraction conditions. More detailed experimental as well as theoretical research is required to clarify this point.

7. Reconstruction-Induced Phase Shifts of RHEED Oscillations

In this chapter, we present a new phenomenon that opens up a new field of in-situ characterization during MBE growth since it allows the real-time measurement of interface quality and segregation. We call this phenomenon the reconstruction-induced phase shift (RIPS) of RHEED oscillations at interfaces. We also demonstrate that the unique properties of the phenomenon shed new light on the current models of RHEED and of RHEED intensity oscillations. The measurements discussed in this chapter are based on a pulsed molecular-beam technique. Here, RHEED intensities are recorded together with reference signals from the shutter motors of the MBE machine at a rate of 0.25 Hz. A convenient way to achieve this is to place LEDs connected to the shutter motors on the RHEED screen within the field of view of the CCD camera. These reference signals then allow the exact synchronization, within 0.08 s, of the onset of different growth pulses typically 20 to 30 s long. The shutters of different effusion cells differ in angular range as well as speed, creating unknown delays of the molecular-beam onsets with respect to their reference signals. These offsets, of about 0.5 s, can be compensated by comparing only growth pulses of the same deposited material and using the same shutter. The technique is then sensitive to the differences between the initial starting surfaces of different pulses as well as the subsequent formation of the interface. The acquisition period of the measurement system is typically 40 to 80 ms, corresponding to one or two full video frames of the CCD camera. This time resolution limits the accuracy of the measurements since it results in an uncertainty of several percent in the phase determination for typical RHEED oscillation periods of 2 s. Note that even changes in growth rate due to the cooling of the cell after shutter opening (shutter transients) are compensated, if the cell returns to its pregrowth state during the growth interruption. To achieve this, the proportional, integral and differential (PID) control parameters of the temperature controllers for the cells were optimized for fast response.

The pulsed technique allows the study of different interfaces with otherwise identical parameters, since growth and diffraction conditions can be accurately kept constant during continuous pulse sequences of several minutes total duration. The growth interruption periods in the sequence are also equal so that the surface returns to a reproducible state at the initiation of

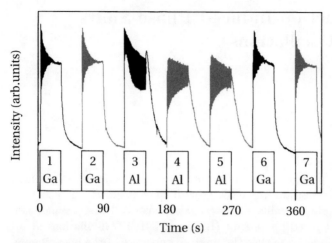

Fig. 7.1. Growth pulse sequence and intensity signal for a continuous timed RHEED intensity measurement. The homoepitaxial reference traces for pulses 3 and 6 are shown in *gray*. The substrate temperature was 590 ± 15 °C with an As$_4$ beam-equivalent pressure (BEP) of 2.7×10^{-2} Pa. The BEP values are to be taken rather qualitatively since the accuracy of the flux monitor was not very high. The intensities were recorded in the [$\bar{1}$10] azimuth. At an incidence angle of 1.3°, the measurement window was placed at approximately 1.9° on the (00) streak, outside the Laue circle

growth. A typical growth pulse sequence for GaAs and AlAs growth is shown in Fig. 7.1. It consists of seven pulses, with pulses 3 and 6 containing the information about heterointerface formation. Pulse 2 (or 7) serves as the homoepitaxial GaAs reference for pulse 6, while pulse 5 is the AlAs reference for 3. Growth interval 4 is needed as a buffer to separate the interface from the reference pulse, and pulse 1 prepares the surface in a defined state for the following sequence. The accuracy of the measurement can be tested by comparing pulses 2 and 7. If they are identical, we can conclude that the growth and diffraction conditions are constant during the measurement sequence and that the growth interruptions are sufficiently long to return the surface to a reproducible state prior to each growth interval.

7.1 Phase Shifts at GaAs/AlAs Heterointerfaces

In Fig. 7.2, the intensity oscillations of the different growth intervals of Fig. 7.1 are shown. The two pairs of traces 3, 5 and 6, 7 start out in phase and end up with a π phase difference. For growth of AlAs on GaAs or GaAs on AlAs, the phase shift in interval 3 or 6 is negative or positive with respect to interval 5 or 7, respectively. There is also a distinct asymmetry in the saturation distance of the phase shift: as indicated by the arrows, the transition at the normal interface (AlAs on GaAs) takes about 11 ML, but for the inverted interface (GaAs on AlAs) it is accommodated in less than 1 ML. In addition, a closer look at the data in Fig. 7.1 also reveals a change in the envelope of the RHEED traces: the completion of the phase shift coincides with the transition towards the decreasing envelope shape of the homoepitaxial growth intervals.

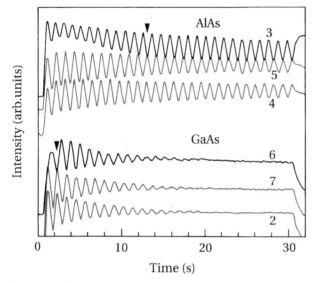

Fig. 7.2. Superposition of the RHEED intensities of the growth intervals of Fig. 7.1. The position of each interval in the growth sequence is indicated by the number to the *right*

Before we turn to the implications of the RIPS effect on RHEED and interface formation, we map out its dependence on As_4 pressure, substrate temperature, growth rate and diffraction conditions to obtain a solid basis for its interpretation.

7.1.1 Variation of As_4 Pressure

Owing to the extended phase shift at the normal interface, this part of the growth sequence is the most sensitive to changes in the experimental parameters. When we vary the As_4 pressure, we obtain the behavior shown in Fig. 7.3. The oscillations are seen to depend drastically on the applied As_4 pressure, with the changes extending out to the boundaries of the 25 ML growth interval. The pressure values are to be taken rather qualitatively since their accurate determination is difficult with the gauges used. At lower As_4 pressures, a 'hump' appears after the growth of about 2 ML. This hump also features an additional shift of the phase, as can be demonstrated in the upper

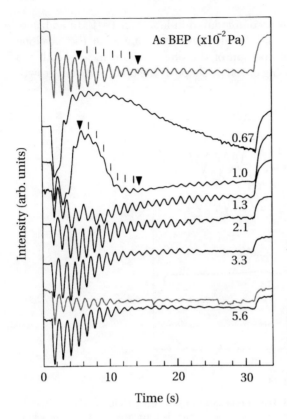

Fig. 7.3. RHEED intensity oscillations from the growth of AlAs on GaAs as a function of As_4 pressure. The sample temperature was $580\,^\circ$C and the oscillations were measured on the specular spot with a beam incidence angle of 0.8° along the $[\bar{1}10]$ azimuth. *Gray curves* indicate the 5.6×10^{-2} and 1×10^{-2} Pa reference curves for growth of AlAs on AlAs. The incident electron energy was $20\,$keV

part of the figure by counting oscillation periods in identical time intervals. Having reached a shift of almost π at the first arrow, the oscillations shift further on the downward slope of the feature, where almost eight periods can be counted at the heterointerface for seven periods of the reference. The transitional hump is only observed at the normal interface. At the inverted interface, the phase shifts back rapidly without exhibiting a significant dependence on the As_4 pressure.

7.1.2 Variation of Substrate Temperature

When we vary the substrate temperature, keeping all other parameters constant, we obtain the behavior shown in Fig. 7.4. Labeling the traces from top to bottom beginning with 1, the measurement sequence was 5, 2, 7, 3, 1, 6, 8, 4 to ensure the absence of drifts between the measurements. Similarly to the As_4 BEP variation, a hump appears for higher substrate temperatures, centered at about 4 ML from the nominal interface. A closer look at a wider range of temperatures, including the reference traces, is contained in Figs. 7.5 and 7.6. Again, the measurements were performed in irregular order to demonstrate the absence of memory effects. The two figures cover a

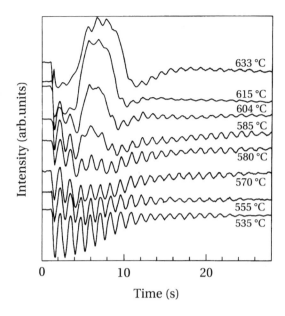

Fig. 7.4. Temperature dependence of RHEED intensity oscillations for AlAs grown on GaAs as a function of the substrate temperature. The detector was placed at the specular spot position for an incidence angle of 0.8° using 20 keV electrons in the [$\bar{1}$10] azimuth. The As$_4$ BEP was 0.13 Pa

range large enough to include the growth conditions generally used for GaAs and AlAs growth. Several regimes can be distinguished. At low temperatures, the RIPS is small and both traces are approximately in phase. Starting at $\approx 520\,°C$, a transitional range follows in which the final shift amounts to less than π. The amplitude of the oscillations increases. From about 550 to 600 °C, a commonly used growth regime for GaAs, the shift is half a period and shows a varying saturation distance, which is plotted in Fig. 7.7. Around 600 °C, the hump feature appears and broadens until it covers almost the entire growth interval. Around the maximum of the hump, the oscillations are almost out of phase, as in the case of As$_4$ pressure variation. For the 609 °C curve, the shift is exactly π at the maximum.

In the region prior to the hump, a small shift to the right with respect to the reference is apparent for the 609 °C trace. For higher temperatures, the oscillations are approximately in phase to the left of the hump until the phase shift of the hump maximum is assumed, almost from the beginning of growth. In the region past the hump, the phase shift is opposite to the value at the maximum, less than π for the 609 °C curve and π at 626 and 643 °C. At higher temperatures, the RIPS saturates at less than half a period. The oscillations for the heterointerface vanish at 702 °C, where GaAs desorbs in significant amounts and degrades the surface morphology that serves as a template for AlAs growth. The oscillations only reappear after the first AlAs pulse, which smooths and stabilizes the surface. AlAs desorption is still insignificant at this temperature, as can be verified by its constant oscillation period in the reference intervals over the entire temperature range.

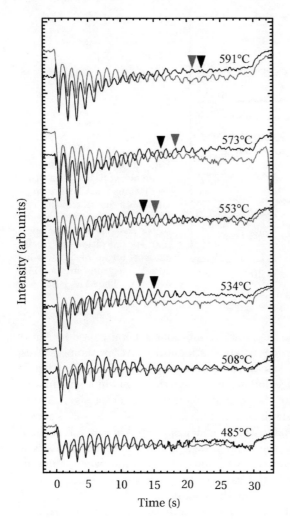

Fig. 7.5. Oscillation traces for growth of (*black*) AlAs on GaAs and (*gray*) AlAs on AlAs as a function of temperature. The measurements were made on the specular spot with 20 keV electrons incident at 0.72° in the [$\bar{1}10$] azimuth

Summarizing the high-temperature behavior, we obtain a shift to the right of $\approx \pi/2$ at the onset and a shift of $\approx 3\pi/2$ to the left at the exit of a well-defined hump centered about 4 ML from the normal interface.

We can verify the symmetry in the saturation value of the RIPS by monitoring its behavior at the inverted interface for the same growth and diffraction conditions. This is shown in Fig. 7.8. For temperatures below 550 °C, the π phase shift is obtained at slightly lower temperatures for this interface compared to the normal interface. In this range, the measurements for the inverted interface are not very reliable beyond the first few oscillations because of the strong damping. Above 550 °C, the phase shifts to an out-of-phase value immediately after growth initiation. Only above ≈ 620 °C does the sat-

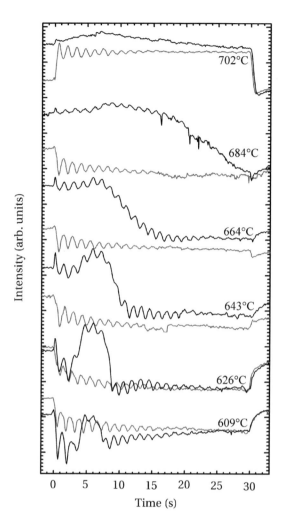

Fig. 7.6. High-temperature part of the RHEED oscillation measurement shown in Fig. 7.5. The sample size in both figures was 1 cm along the beam direction

uration value of the shift become less than π, and at still higher temperatures the oscillations become too weak to reliably determine a phase position.

As general features of the RIPS effect we obtain two remarkable properties. First, the saturation values of the shift, apart from a transitional range at low temperatures, are opposite and of equal magnitude for normal and inverted interfaces probed with identical growth and diffraction conditions. Second, the transitional behavior is drastically asymmetric. Whereas for the normal interface typically 10 to 20 ML are affected, the shift at the inverted heterointerface is typically accommodated within the first ML. This points towards a fundamental difference between these two interfaces inherent to the MBE growth process.

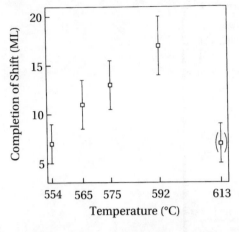

Fig. 7.7. Saturation distance of the RIPS as a function of temperature for the diffraction conditions and As$_4$ pressure used in Fig. 7.1

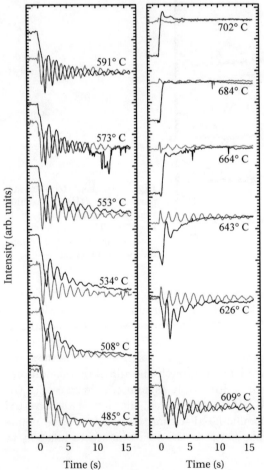

Fig. 7.8. RIPS as a function of temperature at the inverted (GaAs grown on AlAs) interface. The growth and diffraction conditions are identical to the experiments of Figs. 7.6 and 7.7. The *gray curves* represent growth of GaAs on GaAs; *black curves* show the growth of GaAs on AlAs

7.1.3 Variation of the Growth Rate

As the third parameter, we varied the growth rate in a number of otherwise identical experiments. The corresponding curves are shown in Fig. 7.9. It is evident from the figure that growth rate variation at constant As$_4$ BEP has an effect similar to temperature variation or changing the As$_4$ pressure. For higher growth rates, which correspond to a lower As$_4$ flux per ML, the hump starts to emerge, similarly to what is observed for a reduction of the As flux or an increase of the substrate temperature. The only noticeable difference

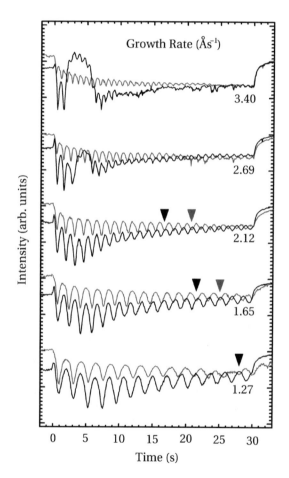

Fig. 7.9. Growth rate dependence of intensity oscillations at the interface. AlAs was grown on top of GaAs with different AlAs growth rates. The sample temperature was 584 °C with a constant As$_4$ BEP of 3.2×10^{-3} Pa. The incidence angle was 0.35° along the [$\bar{1}$10] azimuth on a sample 6×6 mm in size. Again, the *gray curves* are for growth of AlAs on AlAs, whereas the *black curves* are for AlAs on GaAs

is the phase shift prior to the hump, which is π for the highest growth rate instead of ≈ 0 in the temperature variation case. As we shall see in Chap. 9, this can be explained by different diffraction conditions in the experiment. The lowest black arrow marks the saturation distance of the RIPS in the corresponding pair of curves; the other two black arrows mark the same number of oscillations in two pairs of curves for higher growth rates. The actual saturation distance at higher growth rates is larger, indicated by the gray arrows. However, the saturation distance is closer to the black arrows than to a constant transition time, suggesting that the saturation distance in units of an ML does not depend much on the growth rate. This suggests that processes faster than the ones producing the oscillation of the signal are responsible for the transitory behavior of the RIPS. A detailed discussion is presented in Sect. 10.1.

The similarities in the overall dependence of the RIPS effect on As_4 pressure, temperature and growth rate suggest that the As concentration at the sample surface is the common parameter governing the behavior of the effect. Temperature variation also changes the As coverage of the surface in the commonly used As_4 pressure growth mode of MBE. The results therefore point towards changes in the surface reconstruction being responsible for the RIPS effect.

7.1.4 Variation of Diffraction Conditions

The most surprising property of the RIPS effect is the independence of its saturation value of the diffraction conditions. Although the absolute phase of the oscillations is a complicated function of the diffraction conditions (see Fig. 3.5), the relative phase of the two growth intervals in the RIPS is not. Note, for example, that the traces of Figs. 7.3 and 7.5, and 7.8 for 591 °C show the same final value of the shift, although they differ in incidence angle by almost 1.2° and one of them was measured outside the Laue circle. This property holds not only for incidence angle variation, but also for different azimuths as well as for different-order streaks of the same pattern. An example is shown in Fig. 7.10. For the same surface, the RIPS has the same final value independent of diffraction conditions. This indicates that the RIPS is a property of the crystal surface that depends only on the growth conditions. It is most likely not a diffraction-induced effect. Inspection of Fig. 7.10 also reveals that the transitional behavior during heterointerface formation can differ drastically for different diffraction conditions. Whereas a long-range transition is observed for the specular spot on the (00) streak, the phase shifts immediately for the {01} streaks as well as for the specular spot in the [110] azimuth. This variation can be explained by a real-space sensitivity of RHEED and will be treated in detail in Sect. 7.2.4.

7.2 Experimental Results

In this section, we interpret the RIPS in terms of its relation to surface structure and surface properties. These observations lead to important conclusions about the interaction of RHEED electrons with the surface and provide strict criteria to test theoretical models.

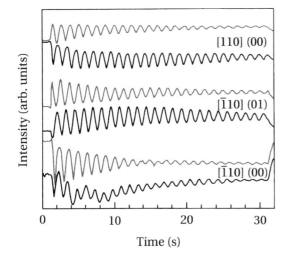

Fig. 7.10. AlAs growth on GaAs (*black*) and homoepitaxial reference (*gray*) for different diffraction conditions. All curves were measured at an incidence angle of approximately 1° with an electron energy of 20 keV. The sample temperature was 580 °C with an As$_4$ BEP of 0.13 Pa. All reflections were measured on the Laue circle

7.2.1 Sampling Depth of RHEED

The extreme asymmetry of the RIPS allows us to define an upper limit for the sampling depth of RHEED. This sampling depth is an important parameter in any theoretical treatment since it is related to the amount of dynamical scattering. The larger the sampling depth, the more complicated a model of RHEED oscillations has to be, to include more layers at the surface and to represent the numerous scattering pathways. As mentioned in Sects. 2.1 and 5.3, the large scattering cross-sections for electrons are responsible for a high degree of multiple scattering inside the crystal. For scattering from surface atoms, kinematical processes are expected to dominate, according to our discussion in Sect. 4.3. We would therefore expect an increasing amount of multiple scattering for large sampling depths, corresponding to large penetration distances into the bulk.

The oscillating contribution to the RHEED signal is expected to originate in the top layer only, since it oscillates with the monolayer period. It could, however, also be influenced by the structure and composition of the underlying bulk, since substrate layers of different conposition could interfere differently with the oscillating beams, leading to phase shifts.

If the RHEED signal contained bulk contributions in this way, the change in signal at a growing interface would be only gradual. Even an abrupt interface crossing the probed volume during growth would generate a gradual transition, with the typical transition distance representing an effective sampling depth of the RHEED electrons. Since the form factors of Al and Ga are significantly different (see Fig. 2.2 or the chemical-contrast TEM images in Figs. 1.2 and 10.3), one would generally expect significant effects from a change between these two materials. The RIPS allows us to directly obtain this transition distance at the interface as the time needed for the transition from alloyed to homoepitaxial growth. It corresponds to the saturation time of the phase shift. Therefore, the saturation time at the inverted interface, where it is smallest, provides us with an upper sampling-depth limit for the diffraction conditions used in the experiment. Typically, this transition distance is below one ML, as can be readily verified by inspection of Figs. 7.11 (arrows), and 7.2 or 7.8.

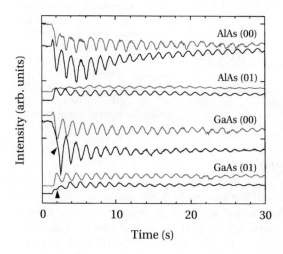

Fig. 7.11. RHEED intensities versus time for a low incidence angle of 0.35°. For all four pairs of traces, the *gray line* is the homoepitaxial reference and the *black line* is the signal from the heterointerface. The {01} intensities can only be measured outside the Laue circle at these low angles. The beam energy was 20 keV with the beam directed along the [$\bar{1}$10] azimuth. The sample temperature was 584 °C and the As_4 BEP was 3.2×10^{-3} Pa. The growth rates were $2 \, \text{Å} \, \text{s}^{-1}$

A second possibility is that bulk contributions are present but do not interfere with the oscillating signal. In this case, the average intensity during growth, with the oscillatory part removed, should show a gradual transition due to the large sampling depth. This means that for measurements where the average intensities during growth are different for GaAs and AlAs, one should observe gradual adjustments of the average intensity at the interfaces. The examples in this work, see Figs. 7.1 and 7.8, demonstrate a different behavior. In many cases, at moderate as well as high temperatures, the new average intensity level is reached during the first monolayer. In Fig. 7.8, this is the case around 580 °C as well as above 670 °C, where the surface is not very well ordered because of the thermal motion of the surface atoms.

This means that the information contained in the RHEED signal and, especially, in its oscillating part is generated in the reconstructed top layer at the surface, at least for low incidence angles. This is contrary to the basic assumption in most dynamical calculations. These calculations, as discussed in Sects. 2.3.2 and 3.2.4, generally involve several MLs of the crystal since they use an exponential damping through the imaginary part of the potential. The very small sampling depths obtained from the saturation distance of the RIPS and the average intensity, however, agree well with our semikinematical simulations of Sect. 4.3. There, we found that generally the top two

lattice planes are sufficient to calculate a meaningful approximation to the
experimental pattern.

7.2.2 Phase Shifts and Surface Reconstructions

From the growth-parameter dependence as well as the low sampling depth,
we expect that the RIPS is intimately linked to the surface reconstruction
and to reconstruction changes at the heterointerface. It is possible to record
the reconstruction profile together with the oscillation information by using
linescans. Instead of integrating the intensity over a two-dimensional area of
the screen, in this measurement mode the intensity profile along a straight line
is recorded as a function of time. The additional spatial information results
in a two-dimensional data set that contains the reconstruction information in
reciprocal-space distances as a function of time. A false-color RHEED pattern
is shown in Fig. 7.12, with the measurement lines marked in white. Two

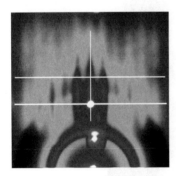

Fig. 7.12. Static RHEED pattern and measurement lines used to record Fig. 7.13. The
sample temperature was 547 °C at an As$_4$
BEP of 3.1×10^{-3} Pa. The data were taken
along the [$\bar{1}$10] azimuth with 20 keV electrons. The circular feature between the horizontal lines is not the Laue circle, but an artifact of the phosphor screen. The incidence
angle is 0.77°. See color plates at the end of
the book

lines were chosen parallel to the shadow edge, one to contain the oscillation
information of the specular spot and the other, at larger angle, to collect
primarily the reconstruction profile. A third line runs along the (00) streak
perpendicular to the shadow edge. The resulting data set for a standard
measurement sequence similar to Fig. 7.1 is reproduced in Fig. 7.13. In the
display, the time axis points upwards, zero coinciding with the beginning of
the first GaAs interval as marked at the right-hand edge of the data. The
reciprocal-space coordinate runs along the bottom edge of the figure and
consists of four different panels, corresponding to the lines in Fig. 7.12. The
leftmost panel represents the upper line, the second panel is the lower line and
the third panel contains the perpendicular line along (00). To the right, the
synchronization data are included in an additional panel for reference. The
rectangular bright areas mark the growth intervals of GaAs and AlAs. The
reconstruction information is most clearly observed in the leftmost data set,
for the line position outside the Laue circle. The GaAs surface reconstructs in
the $\beta(2\times4)$ structure and therefore displays a fourfold profile along the [$\bar{1}$10]
azimuth. For the growth conditions chosen, this periodicity is maintained

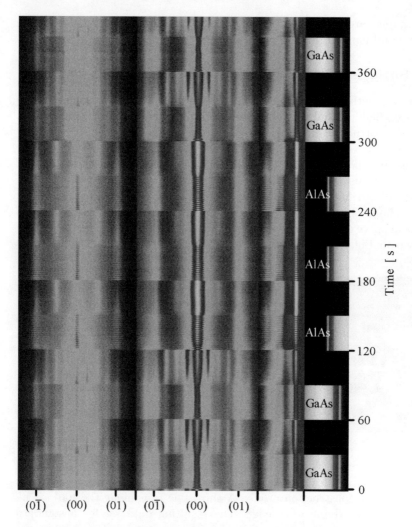

Fig. 7.13. Pulsed RHEED sequence obtained with the diffraction geometry and growth conditions of Fig. 7.12. The changes in the surface reconstruction can be directly observed, most clearly in the *leftmost panel*. The image is divided into four panels, the divisions between the panels being indicated by the *three longer markers* on the horizontal axis. See color plates at the end of the book

during growth. The static AlAs surface shows the $c(4\times4)$ reconstruction with a twofold pattern along the recorded line. During growth, the reconstruction is very weak.

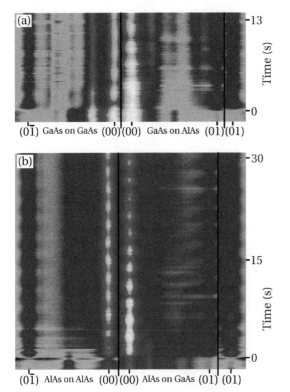

Fig. 7.14. RHEED intensities parallel to the sample surface as a function of time. The sample temperature was 540 °C, with an As$_4$ pressure of 3.2×10^{-3} Pa and 20 keV electrons. See color plates at the end of the book

A closer examination (Fig. 7.14) of the linescans at the heterointerfaces reveals that the asymmetric transition behavior observed for the specular spot is also present on the reconstruction streaks. For the measurement shown in Fig. 7.14, the growth rate was decreased and the As$_4$ pressure was increased so that the $c(4\times4)$ structure of AlAs was present even during growth.

The combination of the three recordings allows direct comparison of the phase on the (00) and {01} streaks. Simultaneously, the evolution of the fractional-order beams, indicating the change in surface reconstruction, can be observed. The RIPS on the (00) streak at the normal interface takes about 7 ML to saturate. This can most clearly be observed by comparing the positions of the maxima of the specular spot (white) in Fig. 7.14a on both sides of the black separation line. In between the (00) and (01) streaks, the fourfold pattern of the GaAs $\beta(2\times4)$ reconstruction disappears simultaneously. This takes about 4 ML, after which the half-order streak of the AlAs $c(4\times4)$ reconstruction starts to emerge halfway between (00) and (01) in the middle panel of Fig. 7.14a. The appearance of the AlAs surface reconstruction coincides with the saturation of the RIPS, indicating a link between reconfiguration of the surface and the phase shift. The inverted interface is shown at the same scale in Fig. 7.14b. The growth rate of GaAs was about the same as for AlAs and all other growth parameters were identical, as Figs. 7.14a,b are taken from the same set of experiments. The RIPS on the (00) streak almost instantaneously reaches its final value and the fourfold reconstruction reappears shortly after the completion of the first layer. The transition distance for the surface reconstruction is of the same magnitude as the transition distance for the RIPS, and it exhibits the same asymmetry at the two heterointerfaces, suggesting a direct connection between the two phenomena. The 2/4 streak is the first to disappear at the normal interface and it reappears last at the inverted interface.

From the simulations of Sect. 4.3 and, especially, Fig. 4.23, we can expect a (2×4) structure with kinks in the dimer rows in both transitional regimes. This can be verified from STM investigations of these interfaces [40]. Figure 7.15 shows the GaAs (2×4) reconstruction prior to normal-interface

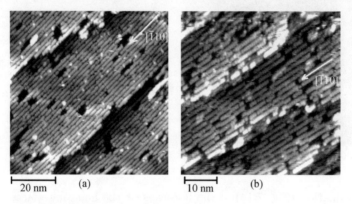

20 nm   (a)   10 nm   (b)

Fig. 7.15. GaAs $\beta(2\times4)$ surface (**a**) and surface structure after the deposition of 10 ML of AlAs at the normal interface (**b**) imaged by STM. The sample temperature was 600 °C with an As$_4$ BEP of 4×10^{-4} Pa [40]

formation and the surface structure after the deposition of 10 ML AlAs. The growth was performed around 600 °C, where AlAs no longer shows the $c(4\times4)$ reconstruction. The conditions are therefore not directly equivalent to Fig. 7.14. Nevertheless, the main features of the RIPS can be verified. An AlAs surface at these growth conditions, corresponding to a (1×1) RHEED pattern, did not permit atomic resolution. Only topographical scans, one of which is shown in Fig. 7.16a, were possible. The presence of a disordered (2×4) structure as in Fig. 7.15b still corresponds to fourfold symmetry in the RHEED pattern. It therefore indicates that the surface is in an intermediate state of interface formation even after 10 ML of AlAs deposition. One

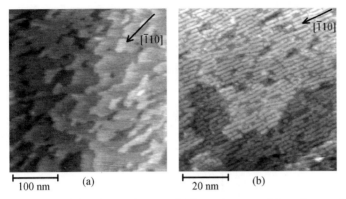

Fig. 7.16. (a) AlAs surface 50 ML from the normal interface and (b) surface structure after the deposition of 1 ML of GaAs on the inverted interface. The sample temperature was 610 °C with an As BEP of 4×10^{-4} Pa [40]

has to take into account with these types of experiment that the surfaces have been annealed for a significant time, since the procedure of quenching to room temperature for the STM experiments takes about 20 min. Therefore the transitory effects at the interface may become strongly overemphasized. Nevertheless, the asymmetry of the surfaces is evident. The inverted interface after 1 ML of GaAs deposition shows a degree of disorder comparable to the normal interface after 10 ML of AlAs deposition, thus directly imaging the different formation process of the two interfaces.

Note that the AlAs surface shows a roughness similar to the GaAs surface [38]. This indicates that the morphological contribution to the diffraction process is not very much different for AlAs and GaAs. A significant reorganization of the growth front in terms of roughness is therefore unlikely to be the cause of the RIPS.

Another way of establishing the connection between the RIPS and the surface reconstruction changes is to correlate the surface phase diagram with the temperature dependence of the RIPS. This is shown in Fig. 7.17 for the experiment of Fig. 7.4. Investigating the surface reconstruction sequence for the static (nongrowing) surfaces as a function of temperature, we obtained the following behavior. For AlAs, the $c(4 \times 4)$ structure vanishes around 600 °C, and an ill-defined transition regime follows. Finally, the surface stabilizes to a (2×4) reconstruction around 650 °C [242]. GaAs loses its (2×4) reconstruction around 625 °C, changing to a diffuse (3×1) pattern.

As we shall show in Sect. 10.1, the transitional range of the RIPS can be linked to Ga segregation at the normal interface. The behavior of the RIPS at the normal interface can therefore be regarded as a sweep through the surface phase diagram in the composition direction. Comparing Fig. 7.17 with Fig. 7.6, we can identify the transitional region of the AlAs reconstruction with the region to the right of the hump where the saturation value of the

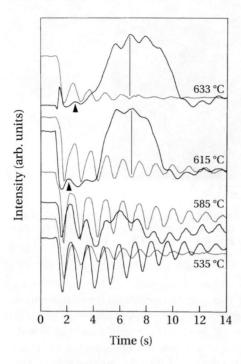

Fig. 7.17. Detailed view of the RIPS at the normal interface for the conditions of Fig. 7.4. The *markers* indicate a phase shift in the prehump region that can be associated with a static GaAs surface reconstruction transition between 615 and 633 °C

RIPS is less than π for higher temperatures. The hump itself is characterized by its own fixed phase relation and two sharp boundaries on both sides, at least for intermediate temperatures. This leads us to conclude that this hump range constitutes a separate transitional surface reconstruction phase. To the left of the hump, in the GaAs regime, the phase jump indicated by the two arrows coincides with the surface phase transition from (2×4) to (3×1). This phase jump does not influence the shape of the following hump feature, indicating that in fact the RIPS is not due to surface kinetics, but is a manifestation of the distinct and mutually independent surface phases.

The hump roughly corresponds to the temperature and composition window of the 'forbidden range', where the surface morphology of the material markedly degrades [243]. We speculate that this surface reconstruction is metal-rich since it is characterized by a strong increase of intensity close to the (200) reflection. This reflection is quasi-forbidden in bulk GaAs because of the similarity of the Ga and As atomic form factors [244] (see also Fig. 2.2) as its intensity is proportional to $16(f_{\mathrm{Al,Ga}} - f_{\mathrm{As}})^2$ [245]. The (200) reflection is therefore used in HRTEM to achieve high chemical contrast. This allows us to associate the drastic increase of intensity in the hump with the presence of Al atoms in the scattering volume. Taking into account the small sampling depth as demonstrated in Sect. 7.2.1, this means that Al must be present at or very close to the surface. Reconstructions with relatively high intensity close to (200) can therefore be assumed to be Al-rich. In this treatment, we always assume that the atoms, even in the reconstruction, are located close to their bulk positions. The assumed connection of the Al content with the intensity of (200) is also supported by the overall behavior of the traces in Fig. 7.6. With increasing temperature, the As content in the reconstruction decreases, a trend generally observed for MBE with an As overpressure. Simultaneously, the intensities in the growth pulse increase until they level out at the peak intensity level of the hump.

We conclude that the experimental evidence points towards a close connection of the RIPS with the structure of the surface reconstruction. The results force us to assume that the changes in the top layer of the surface reconstruction are intimately linked with the relative phase shifts of the RIPS. We therefore describe the effect as *reconstruction-induced*, justifying the assignment of the acronym 'RIPS' at the beginning of this chapter.

7.2.3 Phase Shift Variation Along a Streak

In the next step, we take a closer look at the diffraction condition dependence of the RIPS. In Fig. 7.18, the linescans along the (00) streak of Figs. 7.12 and 7.13 are shown in a combined view. Once again, the asymmetry of the two interfaces is obvious. Similarly to the linescans parallel to the sample surface, the phase shift is completed after roughly 1 ML at the inverted interface. At

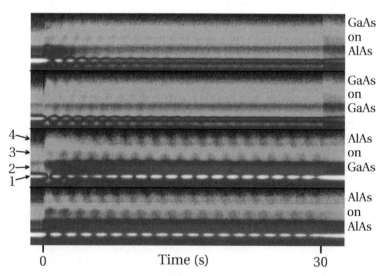

Fig. 7.18. Combined view of linescans along the (00) streak perpendicular to the sample surface. The data are extracted from Fig. 7.13. The scan was recorded along the [$\bar{1}10$] azimuth with 20 keV electrons at an incidence angle (specular spot position) of 0.77°. The sample temperature was 547 °C at an As$_4$ BEP of 3.1×10^{-3} Pa. See color plates at the end of the book

the normal interface, four different oscillating regions along the streak with different exit angles can be distinguished. They are marked and numbered in the second set of linescans from the bottom of Fig. 7.18. An integration over narrow stripes in these different regions yields the diagram in Fig. 7.19. When we compare the different traces, we are left with the astonishing result

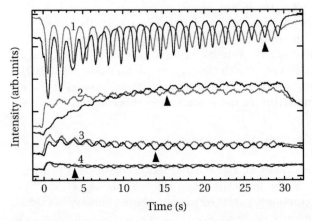

Fig. 7.19. RHEED oscillation traces of the four regions along the (00) streak indicated in Fig. 7.18 (AlAs on GaAs). Although the final saturation values of the RIPS are identical, the approximate saturation distances are markedly different for the different curve pairs

that the saturation distances, marked by the arrowheads, are different for the different oscillating reflections along the streak.

This seemingly contradicts our conclusion from the previous section. If the RIPS is connected to the surface reconstruction, it should show the same behavior everywhere in the reciprocal-space plane of the RHEED screen. A different evolution of the RIPS means that different reflections in reciprocal space monitor different processes in the evolution of the growing surface during heterointerface formation.

We can confirm, however, that the RIPS saturates at the same value for all four regions. This is valid even though the absolute phase positions of, for example, the gray reference traces differ greatly. The identical saturation values, already investigated in Sect. 7.1.4, indicate that all reflections essentially sample the same surface process, since their final value is independent of the diffraction conditions. As we shall see, this paradoxical behavior can be resolved by investigating the phase saturation properties in the direction parallel to the sample surface.

7.2.4 Decoupling of Phase on Different Streaks

An inspection of Figs. 7.10, 7.11 and 7.14 reveals that the extended transition period of the RIPS at the normal interface is only observed on or close to the specular spot in the $[\bar{1}10]$ azimuth. For all other reflections, the phase shift is almost instantaneous. Only at high temperatures is a transition period of a few MLs observed on all streaks at both interfaces, which can be attributed to interdiffusion. The decoupling of the phase along the streak, as well as between different streaks in the same pattern, is only observed in the transitional range at the normal interface. It can be quite drastic. The measurement shown in Fig. 7.20 reveals that in the hump feature, the oscil-

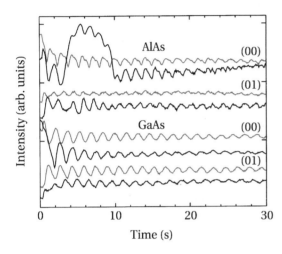

Fig. 7.20. RHEED oscillations with growth and diffraction conditions comparable to those used for Fig. 7.11, but with a reduced As$_4$ BEP of 2.5×10^{-3} Pa. For both AlAs and GaAs deposition, the *gray curves* are from homoepitaxial growth, while the *black curves* are from heterointerface formation

lations in this example are almost out of phase at the specular spot on the (00) streak, whereas they are in phase on the {01} streaks.

An explanation of this phenomenon can be found by reconsidering the kinematical-scattering arguments discussed in Sect. 2.3.1. For an intense reflection on a rod different from (00), the crystal lattice needs to have good long-range periodicity in the lateral direction. For an aperiodic, rough surface, no structure is observed except for the specular spot, which represents the maximum of the autocorrelation function at zero. Keeping in mind the low sampling depth of RHEED of below 1 ML in some cases, a step on the crystal surface already represents a serious disturbance of lateral symmetry.

We can therefore establish a real-space sensitivity of RHEED and RHEED intensity oscillations with respect to surface morphology. The intensity of the higher-order streaks represents terraces on the surface, whereas the intensity of the (00) streak also characterizes areas with no lateral symmetry, such as steps or other defects. Note that in this kinematical consideration the specular spot and (00) streak also contain information from the highly ordered areas of the sample since the autocorrelation function of a perfectly periodic surface also peaks at zero. The argument can be extended towards the variation along the fundamental streak described in the previous section. Again, the reflections away from the specular spot represent parts of the sampled volume that exhibit ordering perpendicular to the surface, whereas disordered contributions are contained in the specular spot. We neglect the disorder scattering plateau that constitutes part of the background, since this background intensity is found not to oscillate measurably.

We therefore find that when we neglect the background, information from the ordered regions of the surface can be extracted from the reflections away from the specular spot, whereas the specular spot also contains contributions from the disordered areas on the surface. This means that the specular spot contains the largest number of different contributions in the RHEED pattern. Therefore, the behavior of the specular spot along high-symmetry directions offers the most complicated starting point for the development of theoretical models. Approaches using dynamical theory to determine the surface structure by fitting rocking curves [137] hence face the additional difficulty of having to take into account an unknown and probably significant amount of perturbation due to disorder.

Using our real-space sensitivity model, we can associate the extended saturation of the RIPS with steps and other disordered features on the surface. The specular spot, in this approach, selectively samples the disordered areas, while information about the well-ordered areas of the sample can be simultaneously obtained from the higher-order reflections.

The disorder due to atoms in the process of moving from site to site does not seem to contribute much to the RHEED pattern, since the streaks remain present during growth. Frequently, reconstruction streaks are visible during growth, even at higher temperatures. This indicates that the atoms spend most of their time at lattice positions. The actual movement from one lattice site to another is fast in comparison and happens on a timescale much shorter than the average dwelling time. This is in agreement with theoretical estimates [155].

Using Fig. 7.21, we propose an explanation for the azimuthal dependence of the saturation distance shown in Fig. 7.10. A typical surface geometry obtained from STM experiments [38] is used, which represents a lower bound for the actual roughness since the annealing times in our experiments were much shorter. Figure 7.21 represents a schematic cross-sectional view along two azimuths of the surface. The scale perpendicular to the surface is expanded by a factor of 100. Time is frozen after the deposition of 0.2 ML of AlAs on GaAs. The filled rectangles represent AlAs islands in the plane of the cross-section, with a mean spacing of about 300 Å. The hatched rectangles schematically indicate islands in the regions behind the plane of the figure. The regions of preferred segregation along the steps are shown by open rectangles. Segregation will be treated in detail in Sect. 10.1. The V-shaped lines depict paths of the RHEED electrons in a ray picture for an incidence

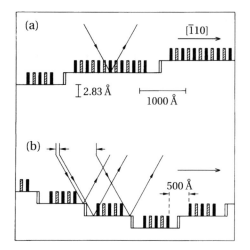

Fig. 7.21. Schematic representation of sections through a GaAs crystal surface. The geometry was obtained from STM experiments. The figure shows the situation after the deposition of 0.2 ML of AlAs. The RHEED beam is able to sample the long and comparatively straight step edges along the $[\bar{1}10]$ direction. In the perpendicular $[110]$ direction, the anisotropy of the islands inhibits the probing of the step edges

and exit angle of 1°. This geometry results in more than 90 % of the 'visible' area consisting of AlAs when we look at the crystal in the $[110]$ direction at the incidence angle of the electrons (Fig. 7.21b). Even the remaining area is not pure GaAs, but $Al_xGa_{1-x}As$ with $x > 0$. This can explain the rapid phase shift along the $[110]$ azimuth, since very soon after growth initiation, the wavefield of the RHEED electrons samples mostly AlAs if its sampling depth is small.

In Fig. 7.21a, the maximum 0.07° miscut of the sample used in Fig. 7.10 produces terraces about 2000 Å wide in the $[\bar{1}10]$ direction. The same argument as before seems to apply, and even more so as the terraces are wider. The difference between the two directions, however, lies in the morphology of the step edges. Steps running along the $[\bar{1}10]$ direction are straight and smooth compared to the ones along $[110]$ because of the anisotropy of the surface features. This anisotropy is large [246], producing more steps perpendicular to $[110]$ than required by a typical vicinality of 0.1° along this direction.

This can also be observed, for example, in Fig. 7.15. Therefore, a large number of relatively long and straight edges along $[\bar{1}10]$ are formed. When growth starts, islands nucleate one mean nucleation distance away from the step edge as atoms closer to the step edge are incorporated at the edge before being able to form island nuclei. Therefore, a channel opens up along the edge where the RHEED beam can reach down to the substrate and sample the step edge. This is not the case for areas away from the edge (indicated by the hatched rectangles), where the islands nucleate in a random pattern [37, 38, 153]. If we assume similar up and down kinetics at the steps, the width of this channel is about 500 Å, as indicated in Fig. 7.21b. This is about the lateral transfer width of standard RHEED systems like the one used in our experiment [122]. On the terraces, the nucleating islands also produce large numbers of steps, but the random nucleation quickly shadows the step edges and very soon diffraction from the top of the forming microterraces can be expected to dominate. This is not the case at the step edges present prior to growth initiation. There, random nucleation is inhibited close to the edge because of the pre-established direction of the step.

8. Energy Loss Spectroscopy During Growth

In a standard RHEED system, the measured intensity is integrated over a range of electron energies, extending from the primary beam energy down to the low-energy cutoff of the RHEED screen of a few keV. The information contained in the electron loss spectrum can be accessed by adding energy loss spectroscopy (ELS) capabilities to the RHEED detector.

Among the several energy loss processes involved in RHEED [78], plasmon losses (10–25 eV) and atomic core-shell excitations (a few hundred to a few thousand eV) are the most easily accessible to in-situ studies during MBE. The typical energy spread of a few eV of a standard RHEED gun is sufficient to study these losses and no modifications to the gun are required.

Several approaches to measuring energy loss spectra during MBE have been realized. Atwater and Ahn mounted a sector-type energy loss analyzer designed for a TEM on a custom-built MBE system to measure core loss intensities during growth [247–251] similarly to techniques used in TEM loss spectrometry [252]. This combination allows the chemical identification of adsorbates on the surface as well as composition analysis with a resolution down to a few percent. The slow acquisition of spectra with the high resolution required, however, makes it less suitable for real-time monitoring of growth.

A different design is the SPA-RHEED, using a channeltron as the detector and a suppressor aperture as the energy filter element [80, 85]. The diffracted electron distribution is scanned across the aperture using electrostatic deflection, allowing the two-dimensional recording of patterns. The energy resolution approaches 4 eV, being limited mainly by the geometry of the suppressor lens. Whereas the spatial resolution of the system is extremely high, the electron-counting detection again limits the temporal resolution of the setup if more than zero-dimensional distributions are recorded. The design of the suppressor lens defines the upper limit of the operating voltage. Values of 6 keV have been obtained [80]. Since As passivates the active surface of the channeltron, the instrument cannot be used in III–V epitaxy with As. The main application of this type of system is the recording of beam profiles from static surfaces, where the high spatial resolution and large dynamic range are important.

The use of apertures and lenses in the detector generally requires a relatively short working distance, exposing the various high-voltage insulators to the deposition or desorption fluxes and usually requiring modifications of the growth chamber.

The disadvantages of channeltron amplification and the problems of coating of complicated electron optics are avoided by using grids as energy filters while retaining the standard phosphor screen as the detector. This approach is used in the standard design of ELS-LEED [253], and similar setups have been developed for transmission electron diffraction [254]. It was adapted to the RHEED case by Horio et al. [255, 256], with some previous attempts by Cowley et al. [257]. With this setup, the working distance is arbitrarily large, allowing easy adaptation to any existing MBE system, with good shielding from the beam fluxes. Using gold grids, chemical reactions with the molecular beams are minimized, thereby solving the As passivation problem of channeltrons. Three spherically shaped grids with a relatively coarse pitch were used by Horio et al. similarly to the LEED setup. The resulting total transmittance of the combined grids was 18 %. Field emission discharge from the charged suppressor grid limited the maximum operating voltage to approximately 5 kV, which is a rather small value for RHEED. Owing to the coarse grids, the spatial resolution of the detector was limited.

To overcome the low operating voltage and spatial-resolution problems of the above design, the author has designed and implemented an improved version of the grid-type detector of Horio et al. The schematic setup is shown in Fig. 8.1. Instead of three spherical grids, two planar Au grids with a period

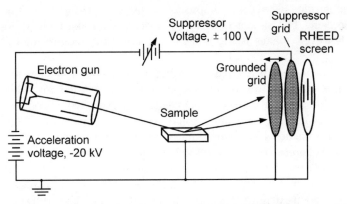

Fig. 8.1. Schematic representation of the ELS-RHEED setup. The RHEED screen and the grounded grid are fixed at zero potential; the suppressor grid is kept at a variable potential close to the corresponding primary-beam energy. All electrons that have suffered an energy loss larger than the potential difference are rejected by the suppressor grid. Differentiation of the intensity with respect to the suppressor-grid potential yields the energy spectrum of the diffracted electrons

of 17 µm and a total transmittance of 20 % are used for the energy filter. The inner diameter of the frames holding the grids is 90 mm, matching the size of a standard phosphor screen. The distance between the grids and the screen is 14 mm, allowing operating voltages of 20 kV by optimizing the shape and geometry of the high-voltage parts. Owing to the large working distance (typically 40 cm), the use of parallel planar grids instead of spherical ones results in negligible distortions of the pattern while sweeping the energy filter. At the same time, the planar geometry allows the use of very-fine-pitch grids, which results in high spatial and energy resolution. A photograph of the detector mounted on a standard MBE chamber is shown in Fig. 8.2.

Fig. 8.2. ELS-RHEED detector mounted on an MBE system. The detector has an arbitrarily large working distance and can be installed in place of the standard RHEED screen on virtually any MBE machine

The detector replaces the standard RHEED screen and no further modifications to the chamber are necessary. Since the electrical design is very simple, with a single electrical feedthrough, the insulators of the suppressor-grid mounting can be effectively shielded against coating, resulting in MBE-compatible uptimes of the detector. The design is sturdy enough to work well even on vibrating cryopumped systems.

Owing to the fringe fields of the capacitor formed by the two grids, the spatial dispersion of the field along the screen is typically several volts. The location with the lowest field strength offers the best energy resolution. To be able to move this point to the area to be studied on the RHEED pattern, the grounded grid can be tilted by means of linear-motion feedthroughs as indicated in Fig. 8.1.

8.1 Electron Loss Spectroscopy on Static Surfaces

The resolution of the ELS-RHEED detector is shown in Fig. 8.3. The measurements were performed on GaAs (001) using a primary-beam energy of 20 keV. Figure 8.3a contains intensity profiles recorded on a circular line along the Laue circle for cutoff energies of 15 and 50 eV. The (00) and three quarter-order peaks are shown, recorded along the $[\bar{1}10]$ azimuth. The spacing between the quarter-order peaks is approximately 1.5 mm on the screen. The inset shows the width of the (00) peak as a function of filter voltage. The peak broadens slightly close to the cutoff. This limits the resolution of the detector to about 1 % of the Brillouin zone for typical materials and chamber geometries. In most cases, this is not a limitation, since typical MBE-grown surfaces are more disordered than the surface of Fig. 8.3a, which was annealed for 3 h prior to the measurement.

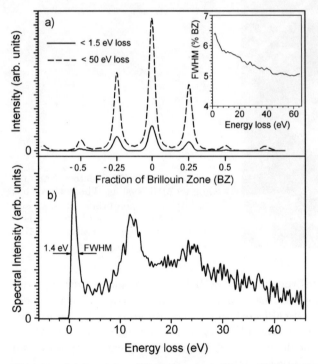

Fig. 8.3. (a) Spatial and (b) energy resolution of the ELS-RHEED detector. Owing to the defocusing effect of the grid close to cutoff, the beam profiles of narrow reflections are widened. This effect, however, is only important for very well-ordered surfaces like the GaAs (001) $\beta(2\times4)$ surface investigated here (annealing time 3 h, $[\bar{1}10]$ azimuth). Profile taken along the Laue circle with an incidence angle of 1.4°. The energy resolution of the ELS-RHEED detector, shown in (b), is below 2 eV. Static GaAs (001) $\beta(2\times4)$ surface, $[\bar{1}10]$ azimuth, incidence angle 1.2°

Energy spectra are obtained by measuring the intensity on the RHEED screen as a function of filter voltage and then differentiating the result. This was done using a low-noise CCD camera and digital image-processing software. A spectrum taken from a GaAs surface is shown in Fig. 8.3b. The spectrum was taken 2 min after growth interruption. The acquisition time was 100 s. Since the noise increases above $\approx 5\,\text{eV}$ energy loss, this currently prohibits the use of the detector for the analysis of core shell excitations. The spectrum consists mainly of surface plasmons at $n \times 11\,\text{eV}$. On well-prepared surfaces, even threefold surface plasmon excitations at 33 eV can be observed. Bulk plasmons at $n \times 16\,\text{eV}$ are hardly detectable. This dominance of multiple surface plasmons is similar to the behavior of liquid-metal surfaces [258] but in contrast to published data on the GaAs surface [259], which show a dominant bulk plasmon peak. We attribute this to the better surface quality achievable in a UHV MBE system compared to a TEM. The fourfold repetition of the same measurement in Fig. 8.4 also demonstrates the excellent reproducibility of the spectra in the low-loss region up to approximately 15 eV. In the spectrum shown, a band-to-band transition peak can be observed about 1.5 eV from the elastic peak.

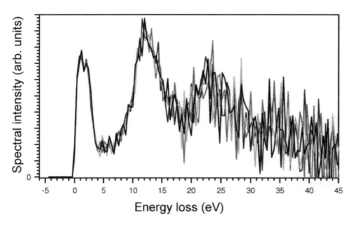

Fig. 8.4. Four repetitions of the same measurement demonstrating the accuracy and reproducibility of the ELS-RHEED detector. GaAs (001) $\beta(2\times4)$ measured on the specular spot in the [$\bar{1}$10] azimuth. The peak at $\approx 1.5\,\text{eV}$ demonstrates the ability to detect interband transitions. The noise increases for higher energy losses

Typically only a small fraction of the total diffracted electron flux has undergone inelastic scattering with losses below 2 eV. The majority of the electrons have excited one or more plasmons. A scan of a larger energy loss region is shown in Fig. 8.5. The 'elastic' contribution to the spectrum is well below 30 % at small incidence angles. Most of the energy losses are in the plasmon region, with an electron–hole continuum tail extending out to several hundred eV. Core losses and other large-energy-transfer events are below the detection limit with our setup. In order to investigate the influence of different inelastic processes on the total RHEED intensity, it is therefore sufficient to study only the low energy loss range, below a few hundred eV loss.

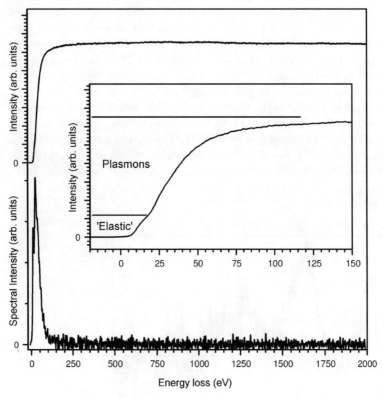

Fig. 8.5. Energy loss spectrum of a GaAs (001) surface measured on the specular spot, $[\bar{1}10]$ azimuth. The majority of the electrons undergo plasmon inelastic scattering; energy loss processes with large energy transfer have a low probability

A series of measurements as a function of incidence angle is shown in Fig. 8.6. In the range below 2 degrees shown in the figure, the relative strength of the various excitations is seen to change dramatically. With decreasing incidence angle, the elastic peak vanishes and all scattering becomes inelastic with energy losses of several eV or more. Multiple plasmon excitations dominate over single plasmon excitations. In the angular range below 1 degree, the character of the diffraction process therefore changes significantly.

The dependence of the plasmon spectrum on surface roughness was investigated by measuring energy loss spectra directly after growth of AlAs on GaAs (001) and comparing them to spectra from the same surface taken 3 min later. The plots of intensity versus filter voltage are shown in Fig. 8.7.

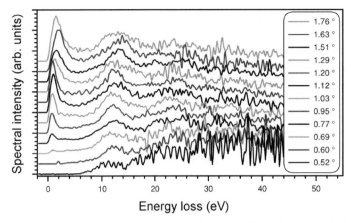

Fig. 8.6. Energy loss spectra as a function of incidence angle from a GaAs (001) $\beta(2\times4)$ surface. The measurements were performed on the specular spot along the $[\bar{1}10]$ azimuth

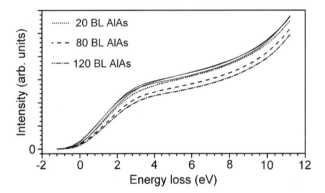

Fig. 8.7. RHEED intensity on the specular spot versus energy loss. Spectra taken immediately after growth (*broken lines*) and three minutes after growth (*full lines*) of AlAs on GaAs (001) are shown. With increasing distance from the interface, less segregating Ga is present on the surface, thus increasing the time constant for surface smoothing. Recording azimuth $[\bar{1}10]$, incidence angle 1.2°. All curves are normalized to the intensity at $\leq 36\,\mathrm{eV}$ energy loss

Since plasmon excitations mainly probe the valence band structure of the material, which is practically identical for AlAs and GaAs, the spectra for GaAs and AlAs are very similar. The thin solid lines show the data taken three minutes after termination of growth. The broken lines represent data taken immediately after growth. Whereas the curves after annealing are practically identical, the curves taken directly after growth are lower in intensity, depending on the distance from the starting GaAs surface. We attribute this to the roughening of the growth front during AlAs deposition. The Al exhibits a lower adatom surface mobility than Ga [260, 261]. Together with the Ga segregation at the normal interface (see Sect. 10.1), this leads to a gradual increase of roughness with distance from the interface during layer growth. Upon annealing, the intensity of all peaks in the spectrum increases. This increase, however, is not evenly distributed between the different inelastic scattering events. All curves are therefore normalized to the intensity at 35 eV loss. A lower position of a curve means that less of the total intensity is due to the elastic and interband processes below 5 eV loss. The lower positions of the nonannealed curves in Fig. 8.7 therefore imply a higher relative excitation probability of plasmon inelastic scattering compared to elastic scattering.

Since the total intensity also increases, the relative increase of plasmon inelastic scattering for rougher surfaces can be attributed to either the decrease of elastic scattering or the increase of plasmon scattering, or both. A decision between the two possibilities is difficult, since most of the total intensity gain is distributed evenly among elastic and plasmon inelastic processes, and the relative variations are comparatively small. In any case, however, the measured difference in the relative excitation probability as a function of surface roughness should allow the in-situ measurement of surface roughness if the method can be calibrated with some independent surface roughness measurement.

A different behavior is observed for the growth of GaAs on AlAs. The data for this case are shown in Fig. 8.8. When overgrowing AlAs with GaAs, a decrease in the surface roughness is expected. This should lead to decreased relative plasmon excitation, resulting in a higher position of the normalized curve. As can be verified from Fig. 8.8, the opposite is the case. After dropping significantly, with a minimum around 11 ML, the intensity recovers and stabilizes at a value slightly lower than its initial position. Segregation can be ruled out as a possible reason for this behavior , since the inverted interface (GaAs on AlAs) is sharp (Sect. 10.1). This means that at a certain thickness of the GaAs layer on top of the AlAs, the relative plasmon excitation reaches a maximum. We assign this behavior to a waveguiding of the surface plasmon wave, with the refractive-index difference between GaAs and AlAs at the buried interface acting as a confinement potential [262].

Materials with a very different band structure can have drastically different electron loss spectra in the region below 30 eV. This can be exploited to achieve chemical sensitivity without having to measure the core-level edges in the spectrum. As an example, we show the spectrum of a β-GaN film grown on GaAs in Fig. 8.9. The GaN layer was 1.2 μm thick and exhibited an n-type carrier concentration of approximately 10^{18} cm^{-3}. The main peak consists of several contributions, indicated by the two shoulders on the high-energy-loss side. Also, the low-energy-loss side of the peak is less steep, indicating a general broadening of the spectral features that can be attributed to the mosaic structure of the crystal, which has a much higher dislocation density than the GaAs. Another broad peak is visible around 20 eV. This is presumably due to a bulk plasmon loss, since the spectrum was recorded on a transmission

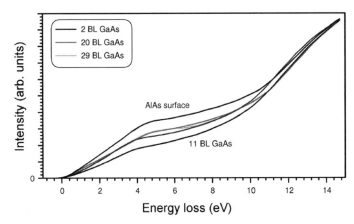

Fig. 8.8. Integrated energy spectra during the overgrowth of AlAs with GaAs. A bilayer (BL) denotes one GaAs molecular layer, usually referred to as monolayer (ML) in the text. A minimum of elastic scattering corresponding to a maximum of plasmon excitation is observed around 11 BL GaAs coverage. The experimental conditions are similar to Fig. 8.7

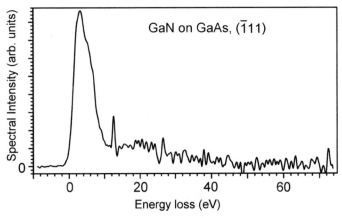

Fig. 8.9. Energy loss spectrum of a 1.2 μm thick cubic GaN film. Apart from a broadening of the peaks, the band-to-band loss region extends up to 8 eV, in distinct contrast to the GaAs surface. The spectrum was taken on a bulk diffraction spot of the RHEED pattern

spot of the RHEED pattern. The sharp feature at 12 eV is most probably an artifact. The clear difference between the spectra of Fig. 8.9 and Fig. 8.3b suggests that ELS-RHEED provides chemical sensitivity that could be used for in-situ and real-time characterization of material composition.

8.2 ELS-RHEED Intensity Oscillations

The biggest advantage of using ELS-RHEED on a fully equipped growth chamber arises from the possibility to study growth dynamics as a function of the electron energy loss [263]. The basic mechanism responsible for the oscillating RHEED intensity during growth is still a matter of debate and Chap. 9 is devoted to this issue. Models have been proposed in which the phase dispersion of the RHEED oscillations as a function of incidence angle is explained by competing elastic and diffuse processes [165, 179, 181, 184]. While elastic diffuse processes do not change the energy loss spectrum, inelastic diffuse scattering should affect the energy loss distribution. In this section, we experimentally investigate the energy loss spectrum during growth to find out how the oscillating intensity during growth affects the energy loss spectrum.

RHEED oscillations from GaAs (001) measured along the [$\bar{1}$10] azimuth are shown in Fig. 8.10. The data in Figs. 8.10a,b are from high-purity vacuum runs; in Fig. 8.10c, N_2 corresponding to approximately one-third of the As_4 background pressure was added, resulting in an inferior surface morphology during growth. In all three panels, the bottom three curves show the raw data with their actual intensity ratio, whereas the top three curves are normalized with respect to the static intensity prior to growth. Typically, only 10–20 % of the total specular-spot intensity at this incidence angle arises from the 'elastic' peak; see Fig. 8.6. Most of the total intensity originates from electrons that have undergone plasmon inelastic scattering. For both the 'inverted' oscillations (oscillating intensity larger than static intensity) in Fig. 8.10a and the 'normal' oscillations (oscillating intensity lower than static intensity) in Fig. 8.10b, the normalized curves are identical within the measurement accuracy. This means that the process causing the oscillating intensity is independent of plasmon excitation. Plasmon inelastic scattering does not influence the shape or phase of the oscillations. Since the momentum transfer of plasmon excitation is largely in the propagation direction, the plasmon energy loss process can be regarded as coherent so far as diffraction is concerned.

Inelastic processes with higher energy losses do not contribute enough to the total intensity to account for a noticeable phase shift (see Fig. 8.5). The remaining processes that could accommodate the diffuse component are elastic diffuse scattering and energy loss processes below the resolution limit of our instrument, notably phonon inelastic scattering. To not show up in the energy spectrum, phonon inelastic scattering and elastic diffuse scattering have to be independent of plasmon inelastic scattering.

The curve in Fig. 8.10c, where N_2 was added during growth, clearly shows that the energy loss spectrum can change during growth. Whether this change is due to roughening of the surface or chemical sensitivity towards nitrogen is not clear at this time. More detailed and systematic studies of altered surfaces are needed to clarify this point.

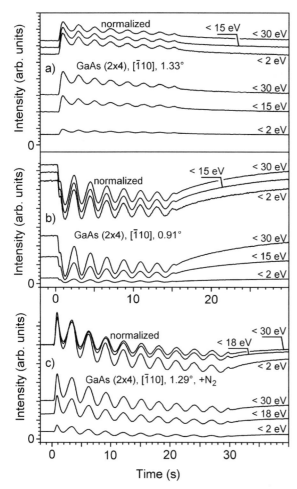

Fig. 8.10. Energy-filtered RHEED intensity oscillations. GaAs (001) homoepitaxy was recorded along [$\bar{1}$10] with different energy loss processes included. In each panel, the *lower curve triplet* shows the measured raw data, whereas the upper triplet is normalized with respect to the static intensity before growth. The curves in (**a**) and (**b**) are shifted for clarity. The presence of N_2, assumed to increase surface roughening during growth, causes the oscillation envelope to shift when plasmon losses are included (**c**). Under clean vacuum conditions, the curves are identical, implying that plasmon losses do not contribute to the mechanism responsible for the oscillations

9. Phase Shifts: Models

In this section, we discuss theoretical approaches to explain the reconstruction-induced phase shifts (RIPSs). The experimental results presented in the previous chapters impose rather strict constraints on a theoretical model because of the richness of phenomena such a model has to describe in a unified approach. At the same time, the values of the potential obtained from the semikinematical simulations of Sect. 4.3 and the Kikuchi line fits of Sect. 5.2 provide independent estimates and checks of the parameter inputs of the theory.

As a starting point, we summarize the main experimental findings for the GaAs/AlAs system:

- The saturation value of the RIPS is independent of diffraction conditions and a function of the growth parameters only.
- The RIPSs are of opposite direction but equal magnitude for the normal and inverted interfaces.
- The RIPS is intimately related to the reconstructed structure of the topmost surface layer.
- The oscillation component from plasmon inelastically scattered electrons is in phase with the component from elastically scattered electrons.
- The absolute phase position of the oscillations depends strongly on the diffraction conditions.
- The saturation behavior is a function of the diffraction conditions. An extended saturation distance is only observed close to the specular spot along the $[\bar{1}10]$ azimuth.

Whereas the last point can be explained by the postulated real-space sensitivity discussed in Sect. 7.2.4, the first five properties of the RIPS require a combined look at the role of surface reconstruction, surface structure and dynamical diffraction in RHEED oscillations.

9.1 Growth-Induced Phase Shifts

We start by examining the saturation value of the RIPS, since it is directly related to the growth. From the experiments of the previous chapters we know that it is independent of the diffraction conditions. At the same time, it is unique to a certain pair of surface reconstructions and must, therefore, like the diffraction-related shifts, be associated with surface structure changes. After the phase shift has saturated, the growth and diffraction conditions in the first overgrowth interval are identical to those for homoepitaxial growth and therefore the heterointerface oscillations should be in phase with the homoepitaxial ones. Since this is not the case, there must be some memory effect of interface formation. The oscillations behave as if a certain amount of material is added or removed at the interface, typically $\frac{1}{2}$ ML for common growth conditions and GaAs/AlAs. At first glance, this is impossible for a constant group III flux as in our experiments. On the other hand, MBE growth for the compounds used in this work is performed under group V element overpressure so that there is always a sufficient supply of group V atoms for the surface to change the structure of its reconstruction.

To examine this idea, we take a closer look at the existing surface structure models for the (001) GaAs and (001) AlAs surfaces. At common As_4 pressures around 560 °C, the RIPS amounts to half a period. Figure 9.1 shows possible models for the (2×4) structure in this range and for the commonly accepted $c(4\times4)$ surface structure. The model of Fig. 9.1c gave the best fit in Sect. 4.3. AlAs (001) reconstructs in the $c(4\times4)$ symmetry shown in Fig. 9.1d. On top of the last complete As monolayer, there are $\frac{3}{4}$ ML Ga and $\frac{1}{2}$ ML As in Fig. 9.1c and $\frac{3}{4}$ ML As in Fig. 9.1d. We propose the following model. For the AlAs reconstruction to form, a complete layer of group III element (marked in gray) needs to be established during the formation of the heterointerface. Since the $\beta(2\times4)$ structure already contains $\frac{3}{4}$ ML, only $\frac{1}{4}$ ML of group III element needs to be added to complete the transition. The additional As is incorporated as soon as the group III layer is completed since it is supplied in excess amounts. Since the group III flux defines the timescale of the experiment, the process of reaching the next completed layer configuration, which corresponds to one RHEED oscillation, is achieved in one-quarter of the time at the normal interface compared to homoepitaxy. Putting it another way, 5 ML of group III element are needed to advance the growth front by 5 (group III + group V) ML in Fig. 9.1d, whereas only 4.25 ML of group III element are sufficient to achieve the same effect starting from the surface shown in Fig. 9.1c and ending in the configuration of Fig. 9.1d. This means that the oscillations have shifted by three-quarters of a period at the heterointerface.

The saturation value of the RIPS is therefore proportional to the difference in group III element content of the two reconstructions referenced to the topmost complete layer.

In the experiments, a value of half a period is obtained. This can be explained by a replacement of As by Al or Ga in the second layer of the $c(4\times4)$ structure. Medium-energy ion-scattering experiments performed on the $c(4\times4)$ GaAs structure determined a coverage of $\frac{3}{8}$ to $\frac{4}{8}$ ML in this layer, indicating that most or all of the sites marked by the divided spheres are occupied by Ga instead of As [264]. These replacements are consistent with

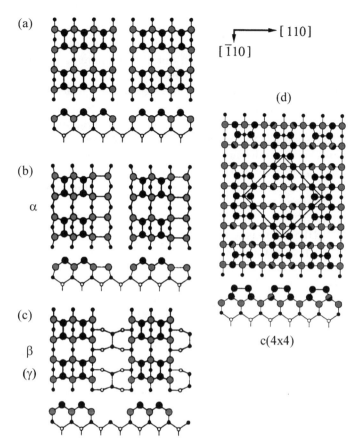

Fig. 9.1. Structure models for ((**a**), (**b**), (**c**)) the (2×4) and (**d**) the $c(4\times4)$ surface reconstructions (adapted from [56]). The top group III layers are marked in *gray*

electron-counting arguments [265]. If half the sites are Ga/Al, the RIPS amounts to half a period as in the experiment, if we start from the Fig. 9.1c configuration. If all the sites are Ga/Al, a half-period RIPS can still be achieved when starting from a structure like that in Fig. 9.1a or 9.1b. The RIPS method cannot therefore provide evidence in favor of any of the three (2×4) models until the As coverage in the second layer of the $c(4\times4)$ structure can be determined unambiguously by an independent experiment. STM studies have so far been unable to determine the composition of the second layer. RHEED experiments to determine the As content of the reconstruction by halo formation at low temperatures indicate an As coverage of 1.25 for the $c(4\times4)$ structure [266]. This would correspond to a 50 % Ga coverage of the second layer and present evidence against the model of Fig. 9.1c.

At present, our method therefore supplies only relative values of the group III element contents of the various reconstructions. The link between a pair of surface reconstructions provided by the RIPS, however, is useful in examining possible surface structure models, since it requires not only the separate consistency of each of the two models, but also compliance with the additional constraint of a fixed group III element relative content.

From Fig. 3.6 we can deduce that within any particular surface reconstruction, the structure and composition of the topmost layers depend also on As_4 pressure. This means that there is the possibility of different group III contents within the same surface reconstruction. The saturation value of the RIPS should be sensitive to such variations. The fact that the saturation value is constant over large temperature intervals, however, indicates that such a dependence is, at most, similar for both reconstructions involved in the transition, if it is present at all.

Revisiting the data of Figs. 7.5, 7.6 and 7.8, we note that for the lowest temperatures the RIPS decreases, indicating that the GaAs surface also approaches the $c(4\times4)$ reconstruction. For temperatures above the $c(4\times4)$ range, no reliable structure models exist for AlAs, making an interpretation in terms of group III element exchange difficult. For all but the highest temperatures in the diagram, the phase shift remains at half a period, indicating a change of $\frac{1}{2}$ ML in the group III content at the heterointerface. At the upper temperature limit of RHEED oscillations, where both reconstructions can be assumed to be group-III-terminated, the shift seems to vanish.

Rapid progress can be expected in the determination of the higher-temperature AlAs reconstructions, since the AlAs phase diagram is about to be established [242] and the first STM scans of pure AlAs surfaces have been obtained (see Fig. 7.16a). The RIPS can be expected to be a useful tool in this process.

9.2 Diffraction-Induced Phase Shifts: The Top-Layer Interference Model

In this section, we investigate models that can explain the dependence of the oscillation phase on the diffraction conditions. We consider incidence angle, azimuth and diffraction order. At the same time, the model should be able to explain phase or amplitude changes due to changing surface reconstructions. Considering the ELS-RHEED results of Sect. 8.2, a model based on kinematical scattering combined with an inelastic incoherent process [183–185] seems unlikely, although the possibility of an elastic incoherent process remains. The fact that the oscillation shape is identical for elastic and inelastic scattering points instead towards a coherent scattering mechanism [145].

9.2.1 A Basic Model

We therefore start by taking a closer look at the top-layer interference model developed by Horio and Ichimiya, presented in Sect. 3.2.5 [192, 267, 268]. This

consists of a one-dimensional solution of the Schrödinger equation perpendicular to the surface, obtained by matching the wave function and its derivative
across the potential discontinuities between layers. To reduce the number of
free parameters and to avoid the pitfalls of birth–death models discussed
previously, we use a perfect layer-by-layer growth model. This means that
growth starts from a flat surface and that one layer is completely finished
before the next one is begun. The coverage and therefore the potential in the
growing layer increase linearly with time, from zero to the value of the underlying material. Mathematically, this corresponds to the textbook example
of a quantum mechanical particle incident on a twofold downward potential
step.

Analogously to [192], the reflectance can be calculated as the square of
the ratio of the reflected wave amplitude B to the incident wave amplitude
A:

$$R = \left(\frac{B}{A}\right)^2 = \left(\frac{K^- a_{\text{lay}} \exp\left(-iK^+ d\right) + K^+ b_{\text{lay}} \exp\left(-iK^- d\right)}{K^+ a_{\text{lay}} \exp\left(iK^- d\right) + K^- b_{\text{lay}} \exp\left(iK^+ d\right)}\right)^2, \quad (9.1)$$

where $K^+ = K_\perp + k_{\perp\text{lay}}$ and $K^- = K_\perp + k_{\perp\text{lay}}$. The amplitudes a_{lay} and
b_{lay} of the downward and upward waves within the layer are given by

$$a_{\text{lay}} = \frac{k_{\perp\text{lay}} + k_{\perp\text{sub}}}{2k_{\perp\text{lay}}} a_{\text{sub}} \quad (9.2)$$

and

$$b_{\text{lay}} = \frac{k_{\perp\text{lay}} - k_{\perp\text{sub}}}{2k_{\perp\text{lay}}} a_{\text{sub}}. \quad (9.3)$$

The surface-normal component of the wavevector in vacuum is denoted by
$K_\perp$. The surface-normal components of the wavevectors in the layer, $k_{\perp\text{lay}}$,
and the substrate, $k_{\perp\text{sub}}$, are given by

$$k_{\perp\text{lay}} = \sqrt{K_\perp^2 + U_{\text{lay}}} \quad (9.4)$$

and

$$k_{\perp\text{sub}} = \sqrt{K_\perp^2 + U_{\text{sub}}}, \quad (9.5)$$

with

$$U_{\text{lay}} = \frac{2me\theta V}{\hbar^2} \quad (9.6)$$

and

$$U_{\text{sub}} = \frac{2meV}{\hbar^2}, \quad (9.7)$$

where m is the electron mass, e the electronic charge, θ the coverage of the
layer and V the average crystal potential.

The oscillations obtained from this model are plotted as a function of
incidence angle in Fig. 9.2. The layer thickness is 1 ML (2.82 Å), and for

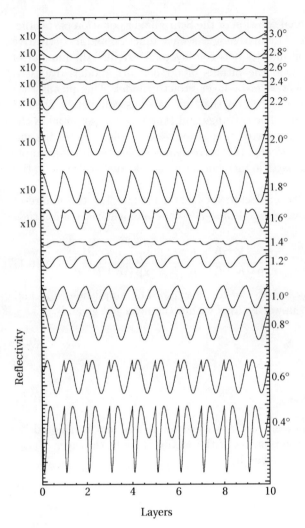

Fig. 9.2. RHEED oscillations for perfect layer-by-layer growth using the top-layer interference model. The layer thickness is 2.82 Å with a 10 V layer potential and an electron energy of 20 keV. Absorption is set to zero

the potential a value of 10 V obtained from the Kikuchi line fits is used. Strong variations in the phase and shape of the oscillations can be observed. All simulated traces show intensity maxima at integer coverages. This is due to the use of a perfect layer-by-layer growth model. The reflectance is always largest for a single interface configuration. As soon as the second interface is added, the interference in the layer reduces the reflectivity except for constructive interference of the two beams. This spike, which is usually absent in experimental data, can be removed in the simulations by including more layers, at the expense of complicating the interpretation because of the added effects from the growth model chosen. Oscillation shapes similar to the 0.6° curve in Fig. 9.2 have been experimentally observed in the first oscillation of GaAs grown on GaAs after long-time annealing [269]. The experiments were made at angles below 1° using an incident electron energy of 15 keV. In this case, the experimental situation more closely resembles our model of a perfect layer-by-layer growth starting from a perfectly flat surface.

The amplitude of the oscillations decreases strongly with increasing incidence angle, a feature frequently observed in experiments. It exhibits local minima at around 0.5°, 1.5° and 2.5° in Fig. 9.2. At these local minima, the phase (defined as the position of the minimum within one oscillation) of the oscillations jumps by approximately $\pi/2$. An overview of the amplitude and phase is plotted as a two-dimensional function of both the incidence angle and layer potential in Fig. 9.3. A comparison with Fig. 3.12 reveals that at the

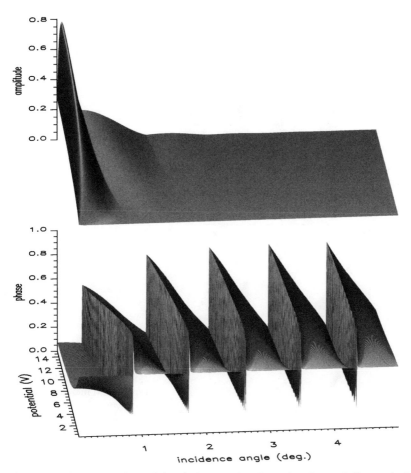

Fig. 9.3. Quasi-3D plots of (*top*) the simulated amplitudes and (*bottom*) the simulated phase positions for $d = 2.82\,\text{Å}$ and varying layer potential V as a function of incidence angle. The electron energy was fixed at $20\,\text{keV}$

phase jump positions both beams are in phase at half-layer coverage, introducing an additional maximum in each oscillation. This results in two minima, one of which decreases while the other increases as the incidence angle increases. In between these phase jumps, the position of the minimum changes gradually. For typical potentials, the phase shows a sawtooth-shaped behavior as a function of incidence angle. Note that this model is a straightforward extension of the kinematical model of Sect. 3.2.2. For very small potentials the phase stays constant at the expected value of 0.5 (minimum at half-layer coverage), with dispersion only in increasingly narrow ranges around the in-phase positions. These in-phase positions are then evenly spaced and located at the positions of the Bragg peaks an X-ray experiment would produce. The amplitude also goes to zero for small potentials since the model does not include any scatterers at the interfaces, and for small potential differences the interfaces become increasingly transparent.

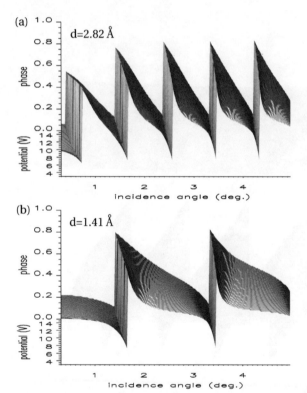

Fig. 9.4. RHEED oscillation phase versus layer potential and incidence angle for layer thicknesses of (**a**) 2.82 and (**b**) 1.41 Å. The electron energy is 20 keV

Figure 9.4 shows two plots of the top-layer interference model along the potential axis. Whereas Fig. 9.4a was simulated for a layer thickness of 2.82 Å, the bulk (002) lattice spacing for GaAs, Fig. 9.4b shows a simulation for 1.41 Å. In both cases, a change in potential causes the position of the phase jumps to shift down towards lower angles with increasing potential. This is the same phenomenon as we have already encountered in the distortion of the Kikuchi line pattern in Sect. 5.2. The variation in the positions is small, except for very low angles, which means that the value of the potential is not a very sensitive parameter in fitting the theory to the experimental results. The Bragg conditions (in-phase conditions) become potential-dependent and, in the context of the top-layer interference model, we therefore denote them as 'generalized Bragg conditions', since they are shifted from their kinematical values although they still represent the conditions for constructive interference. By comparing Figs. 9.4a and 9.4b, it becomes obvious that the model depends much more sensitively on the layer thickness d. Variation of this parameter basically compresses or expands the data along the incidence-angle axis. Since the phase jump positions can be determined experimentally, this should allow accurate fits to experimental data. Most dynamical RHEED calculations include inelastic scattering by introducing an imaginary part of the potential that damps the amplitude of the wave function. A simulation with various values of the imaginary part is shown in Fig. 9.5. From the data,

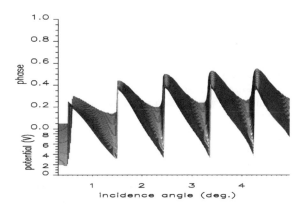

Fig. 9.5. Position of the RHEED oscillation minimum with absorption included in the top-layer interference model. The layer thickness is 2.82 Å, with 20 keV electrons. The real part of the layer potential is kept at 10 V with the imaginary part varying from 0 to 9 V

we can easily verify that the introduction of an imaginary part of the potential does not change the positions of the generalized Bragg conditions. The main change is a reduction in the amplitude of the phase variations with a reduction in the phase discontinuity at the phase jumps. This means that the *positions* of the phase jumps are reliable indicators of the interference layer thickness in RHEED experiments. At the same time, the absorption in the layer modeled by the imaginary part of the potential provides an independent fitting parameter for the amplitude of the phase variations. It turns out, however, that a simulation without absorption agrees best with the experimental data. We therefore neglect absorption for the remainder of this discussion.

9.2.2 Comparison With Experiments

We compare the results of our simulations with the experimental data set of Zhang et al. [167] reproduced in Fig. 3.5. Since their data was taken at 12.5 keV, an appropriately adjusted simulation is superimposed on the experimental data in Fig. 9.6. At first glance, there is no clear coincidence. The

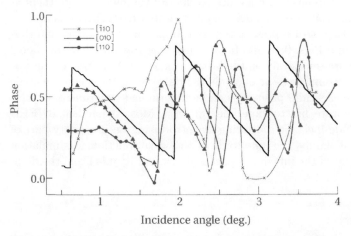

Fig. 9.6. Simulated RHEED oscillation phase (*black line*) for incident electron energies of 12.5 keV and experimental data from Fig. 3.5. The layer thickness used in the calculation is 2.82 Å, with no absorption and 10 V for the real part of the layer potential. The parameters of the simulation were chosen to match the experimental conditions

fit is worst for the [1̄10] azimuth. This is not too surprising, considering the results of Sect. 4.1, which suggest that diffraction is mostly kinematical along high-symmetry azimuths of well-ordered surfaces. In the sequence [1̄10], [110], [010], the fit becomes increasingly better with decreasing lateral symmetry of the reconstruction. The data for the [010] azimuth already shows a fairly clear sawtooth shape, although the period, and therefore the layer thickness of the simulation, is off. This behavior suggests that the agreement of the top-layer interference model with experiment becomes increasingly better as one uses lower-symmetry azimuths to do the measurement. This is not astonishing, since our model totally neglects any lateral structure of the surface and should therefore be best along the 'lowest-symmetry azimuth', also called the one-beam condition [192]. In this approximation, all influences from lateral symmetry are neglected.

In the experiment, the one-beam condition corresponds to a RHEED pattern with a minimum number of Kikuchi lines crossing the (00) streak, indicating the largest possible distance from a crystal zone axis according to our treatment in Chap. 5. In a different investigation [270], another phase jump at 0.68° in the [130] (lower symmetry) azimuth is observed that agrees with the simulation in Fig. 9.6 and further supports our assumption that the model should work best in the one-beam condition.

We therefore recorded RHEED oscillations as a function of incidence angle along the $[\bar{2}10]$ azimuth. The experimental curves for incidence angles between approximately 0.25 and 1.75 degrees are shown in Fig. 9.7. The data were recorded without changing the sensitivity of the detector, to represent

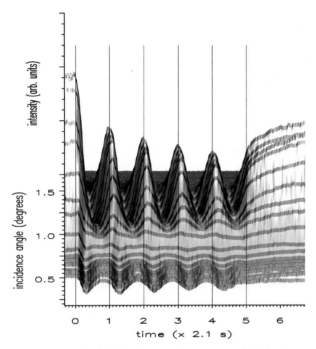

Fig. 9.7. GaAs RHEED oscillations recorded during GaAs homoepitaxy in the (2×4) surface reconstruction as a function of incidence angle, plotted as a quasi-3D surface. The oscillations were recorded at the specular spot position. The oscillation periods corresponding to the deposition of one layer are marked by *vertical lines*

the different absolute intensities at different incidence angles. A comparison with the curves in Fig. 9.2 reveals that the top-layer interference model reproduces all important features of the experiment. The oscillation minimum is located at the left-hand side of the period for low angles, then changes to the right at the generalized Bragg condition crossover, with low oscillation amplitude. Subsequently, the minimum again moves to the left, combined with an increase of the oscillation amplitude. A fit of the top-layer interference model to the experimental data is shown in Fig. 9.8. In this experiment, the determination of the incidence angle could not be performed with great accuracy. In addition, the data do not include the second generalized Bragg condition. The best-fit parameters of $V = 11\,\mathrm{V}$ and $d = 3\,\text{Å}$ are therefore not very reli-

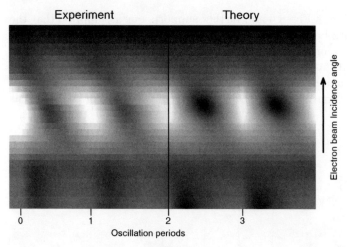

Fig. 9.8. Fit of the top-layer interference model (periods 3 and 4) to the experimental data of Fig. 9.7 with damping removed by a linearly increasing factor (periods 1 and 2)

able. Nevertheless, the good matching of the curve shapes indicates that the top-layer interference model is a valid approximation. All main features of the experiment are reproduced in the fit.

In the fitting procedure, the theoretical reflectivities were multiplied by a fitting factor depending on the incidence angle to match the experimentally observed intensities. For incidence angles to both the low and the high end of the range shown, the experimental intensities are below the values expected from the theory. At low angles, we attribute this to a combination of several factors. The sample size in this experiment was 4×4 mm to reduce nonuniformity damping (Sect. 3.1). This results in part of the beam missing the sample surface at low angles, not contributing to the total diffracted intensity. The onset of this reduction in intensity would imply a beam diameter of ≈ 100 μm, which is a typical value for the RHEED gun [85] used in this experiment. In addition, the strong increase in plasmon inelastic scattering (see Fig. 8.6) might contribute to this reduction, as well as the mesoscopic undulation of the surface [271], shadowing parts of the surface from the incident electrons at small angles. The origin of the reduction in intensity at larger angles is unclear at present.

An experimental data set covering a larger range of incidence angles is shown in Fig. 9.9. In this measurement, the sensitivity of the detector was adjusted for minimum noise levels. The relative intensities of curves at different incidence angles are therefore not exactly defined. The minima of the oscillations continuously shift to the left, covering approximately $2\frac{1}{2}$ periods of the top-layer interference model. This allows us to use the position and spacing of the generalized Bragg conditions (phase jumps) as the most reliable points

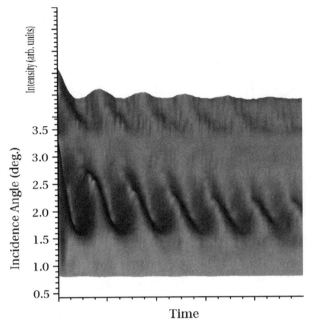

Fig. 9.9. RHEED intensity oscillations recorded in the $[\bar{2}10]$ azimuth of a (001) GaAs (2×4) surface similar to that shown in Fig. 9.7. The angular range includes more than one generalized Bragg condition

for the adaptation of the model. A measurement where the incidence angle was determined with high precision is shown in Fig. 9.10a. For the fitting of the model, the value of the potential of 10.5 V obtained from the Kikuchi line fits in Sect. 5.2 was used, which leaves the layer thickness d as the only remaining fitting parameter in the top-layer interference model. The theoretical curve was calculated with $d = 2.4\,\text{Å}$. This value is less than the bulk (002) lattice spacing of $2.82\,\text{Å}$, which would be expected for interference in the growing bulk structure layer.

We have to take into account, however, that the growing surface is reconstructed. To identify the layer that the interference model describes, the experiment of Fig. 9.10a was repeated with AlAs homoepitaxy in Fig. 9.10b. Since the lattice mismatch between GaAs and AlAs is well below 0.1 %, the experiment would reproduce the GaAs result if the interference were to take place in a bulk structure layer. As can be verified from Fig. 9.10b, however, the spacing of the generalized Bragg conditions is distinctly different. The best fit in Fig. 9.10b was obtained for $d = 3.8\,\text{Å}$. This clearly indicates that the surface reconstruction on the growing layer has to be taken into account in the analysis. The layer thicknesses obtained for both GaAs and AlAs suggest that the interference takes place in the reconstructed layer on top of the growing bulk structure layer as indicated in Fig. 9.10c. Comparing the fit

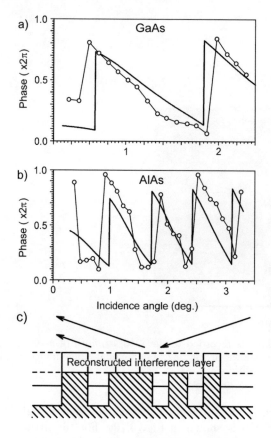

Fig. 9.10. Comparison of experimental data recorded along the [$\bar{2}$10] azimuth for (**a**) GaAs (001) $\beta(2\times4)$ and (**b**) AlAs (001) $c(4\times4)$. The layer-thickness fits (*thick lines*) with a potential of 10.5 V use layer thicknesses of (**a**) 2.4 Å and (**b**) 3.8 Å. The resulting model is shown in (**c**), where the interference takes place in the reconstructed layer on top of the growing layer

values with the structure models of Fig. 9.1, the layer thicknesses determined agree remarkably well with the structural data of the $\beta(2\times4)$ and $c(4\times4)$ surface reconstructions, which consist of one and one-and-a-half atomic bilayers, respectively. In both cases, the reconstructed layers are incomplete and relaxed towards the bulk, presumably reducing their effective thicknesses.

Comparisons of the curve shapes obtained from the top-layer interference model with the experimental data are given in Figs. 9.11 and 9.12. Note that the curve shape, for example in the 0.59° curve of Fig. 9.11, shows a more modulated shape at growth initiation. As growth progresses, the higher-order Fourier components in the signal gradually decrease. This supports our assumption that the absence of the model's sharp peak in the experiment is due to the growth front spreading out over more and more levels during growth. Quite possibly, an extrapolation of the experimental curve towards the left would lead to an even better match with the theory. Since actual growth can never be perfectly layer-by-layer, the spike predicted by the model is always reduced. This assumption is supported by the AlAs data in Fig. 9.12, which generally show a more sinusoidal shape. This is due to the lower surface

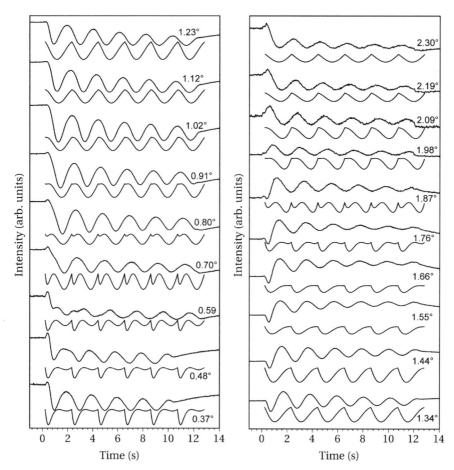

Fig. 9.11. Comparison of theoretical and experimental curves for GaAs data shown in Fig. 9.10a

mobility of the AlAs units, leading to a stronger deviation from the perfect layer-by-layer growth.

If we take this into account and neglect the sharp spike of the theory, the agreement between the theoretical and experimental curve shapes is excellent, considering the simplicity of the model. Except for the low-angle range of the AlAs data, the theory closely follows the experimental data over the entire sampled range. We can therefore be quite confident that our model represents the basic mechanism responsible for the RHEED intensity oscillations.

Quite generally, the approximation of a semiconductor surface by a reconstructed layer on top of bulk material seems a promising approach. The simplest model of a surface is just an interface between homogeneous material and air or vacuum. Improving this to a model that includes more structure, it seems more adequate to go to a configuration that has one layer with dif-

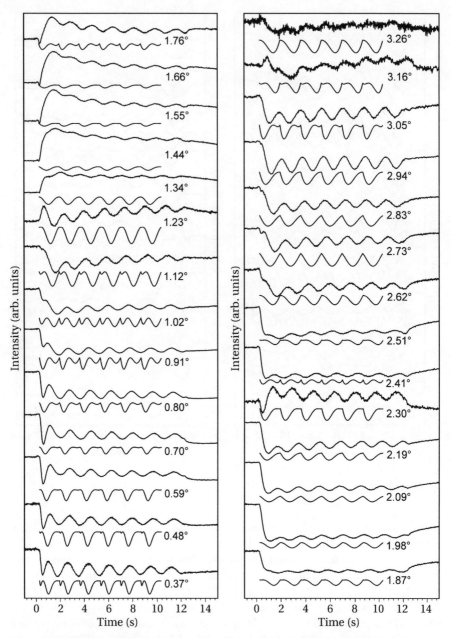

Fig. 9.12. Comparison of theoretical and experimental curves for AlAs data shown in Fig. 9.10b

ferent properties on top of the bulk material instead of dividing the material into identical (crystalline) layers all the way up to the interface.

In our case, the presence of a different layer on top of the bulk material also allows diffraction in the reconstructed layer to be destructive even for static (nongrowing) surfaces. This can provide a qualitative explanation for the 'inverted' curve shapes (compare, for example, Fig. 8.10b with Fig. 8.10a), in which the RHEED intensity during growth is higher than from a nongrowing surface. When we take a closer look at Figs. 9.11 and 9.12, it becomes clear that these inverted oscillations roughly coincide with the positions of the generalized Bragg conditions. These are characterized by constructive interference of the two beams around half-layer coverage, which in turn means destructive interference around integer coverages for realistic systems. This minimum in the reflectance is most pronounced for a more perfect surface during growth interruption, which leads to an increase in the intensity during growth, when the surface is less ordered.

We can summarize the results of this section as follows [238, 239]: *The bulk plane spacing determines the period of RHEED intensity oscillations; the surface reconstruction during growth determines the phase of the oscillations.*

Note that the amplitude of the measured phase variations in Fig. 9.10 is not smaller than the calculated amplitudes, justifying our neglect of absorption in the model in view of the simulations in Fig. 9.5.

9.2.3 Phase Shifts at Interfaces

At this point, our model already allows us to simulate phase shifts at heterointerfaces. In particular, close to the generalized Bragg conditions, huge variations in the phase can be achieved by small changes in the input parameters. Figs. 9.13 and 9.14 show two examples, one for a change in layer thickness and the other for a varying layer potential. The parameter is linearly varied during the first 10 ML and then kept constant. In both graphs, reference curves using the final parameter set are added in the top part of the figure. Since after the adjustment the parameters are identical for both curves, there is no resulting phase shift. The curves, however, show the initial phase shift familiar from the experiments of Sect. 7.2.3. The model shows us why the initial phase offset is not independent of diffraction conditions like the saturation value of the RIPS. For additional examples of the initial phase shift, compare for example the first oscillations of the 609 °C and 626 °C curves of Fig. 7.6 with the corresponding 615 and 633 °C curves of Fig. 7.17, or the initial phase shift variations along the (00) streak in Figs. 7.18 and 7.19. When we add the (saturation value) phase shift due to release or incorporation of a group III element as discussed in Sect. 9.1, we can account now for all observed phenomena. The initial offset due to different layer parameters is accompanied by the possible exchange of group III material, resulting in the combination of a diffraction-condition-dependent initial phase shift

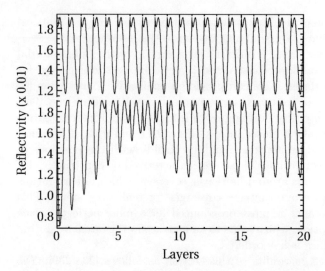

Fig. 9.13. RHEED oscillations at an incidence angle of 1.5° simulated for 20 keV electrons and a layer potential of 10 V without absorption. In the *lower curve*, the layer thickness was increased from 2.5 to 3 Å during the first 10 ML, whereas it was kept at 3 Å in the *top curve*

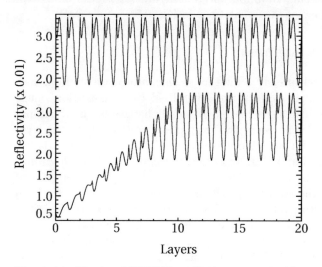

Fig. 9.14. Simulated RHEED oscillations at an incidence angle of 1.5° for a layer thickness of 2.82 Å and 20 keV electrons. Absorption is zero. During the first 10 ML of the *lower curve*, the layer potential is ramped from 5 V to 15 V. In the *top curve*, the potential is kept at 15 V

with a diffraction-condition-independent saturation value of the phase shift (Fig. 7.19).

Transitions through ranges of doubled oscillations similar to those seen in the simulations can be observed for growth of AlAs on GaAs at the normal interface at the appropriate incidence angles. A nice example is shown in Fig. 10.12, where the surface reconstruction was modified by the predeposition of Sn prior to the growth sequence. Note that the absolute intensity during this transition process does not change very much, while the phase varies strongly. This is closer to the simulated situation of Fig. 9.13 and suggests that the changing thickness of the reconstructed layer is the dominant factor in this transition. The layer potential in Fig. 9.14 needs to be changed by a factor of three to observe a phase shift similar to that produced by a layer thickness variation of 20%. At the same time, the absolute intensity in Fig. 9.14 changes by a factor of almost six in this process. The higher probability of a thickness change is in agreement with the Kikuchi pattern simulations that gave the same values of the potential for both GaAs and AlAs surfaces and also supports our interpretation in Fig. 9.10 of a changing reconstruction-layer thickness. A comparison of the fits in Fig. 9.10 with the simulations in Fig. 9.14 clearly reveals the impossibility of fitting both Fig. 9.10a and Fig. 9.10b with the same layer thickness by varying only the layer *potentials* within a reasonable range.

10. Applications
of Reconstruction-Induced Phase Shifts

After having analyzed the basic processes governing the RIPS, we apply it in this chapter to the characterization of segregation as well as to surfaces and surface phenomena different from GaAs/AlAs (001). The examples are intended to show how the RIPS can be used to study interface formation in MBE. An adaptation of the methods to related problems or materials systems should be straightforward.

10.1 Ga Segregation at AlAs/GaAs Interfaces

Ga segregation has been studied extensively by different techniques [24, 27, 51, 53, 54, 272–274]. In this section, we demonstrate that we can detect Ga segregation in situ during crystal growth using the RIPS [200, 275].

10.1.1 (001) Interfaces

Since the surface reconstruction depends on alloy composition and the RIPS monitors changes in the reconstruction, the RIPS can be used to measure the composition changes at heterointerfaces. To obtain the dependence of the phase shift on alloy composition, the AlAs pulses of the measurement sequence shown in Fig. 7.1 were replaced by $Al_xGa_{1-x}As$ with different x in a series of measurements. The results are shown in Fig. 10.1. The measurements were performed at a sample temperature of $580\,°C$ with an As_4 BEP of $1.3{\times}10^{-2}\,Pa$, and the growth rates were adjusted to $2.5\,Å\,s^{-1}$ $Al_xGa_{1-x}As$ for all x used. The intensity traces were recorded at the specular spot position and on the first-order diffraction streak at the Laue circle. Because of the rapid phase shift, the inverted interface was chosen for the measurement to increase accuracy. The saturation values of the RIPS were again checked to be of equal magnitude but opposite direction for any fixed x. The phase shifts are the same on both the (00) and (01) streaks, as shown in Fig. 10.1b. The empirical relationship obtained from the oscillation curves is plotted in Fig. 10.1c. Its nonlinearity allows a sensitive detection of small Ga fractions on the surface of AlAs.

Applying the relationship of Fig. 10.1c to the normal interface, we obtain the segregation profile marked with the filled circles in Fig. 10.2. The growth

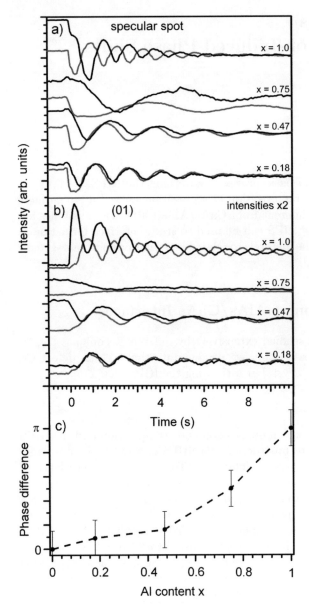

Fig. 10.1. Correspondence between alloy composition x and RIPS saturation value. $Al_xGa_{1-x}As$ was grown instead of AlAs in the measurement sequence of Fig. 7.1 to obtain the relationship plotted in (**c**). The GaAs on GaAs (*gray*) and GaAs on $Al_xGa_{1-x}As$ (*black*) curves at the inverted interface for (**a**) the (00) and (**b**) the (01) azimuth are shown

and diffraction conditions were identical to the calibration measurement with an AlAs growth rate of $1.9\,\text{Å}\,\text{s}^{-1}$. Ga segregates strongly on the AlAs surface. This in turn means that the calibration measurements of Fig. 10.1 contain systematic errors. The surface must be Ga-rich compared to the bulk. This does not render the results flawed, provided that the ratio of top-layer composition to subsurface composition remains the same during the oscillation experiment at the normal heterointerface. We have already seen in Sect. 7.1.3 that the transition behavior of the RIPS does not depend much on the growth rate. This indicates that the segregation process takes place faster than the typical timescales involved in island nucleation and coalescence. We can therefore expect that the difference in the ratio of the bulk to the surface composition between the calibration measurement of Fig. 10.1 and the measurement of the segregation profile (Fig. 10.2) is not too large. The approach to equilibrium for segregation should be faster than a typical oscillation period. The additional 30 s annealing time in the calibration measurement can therefore be expected not to change the ratio of the bulk to the surface composition excessively.

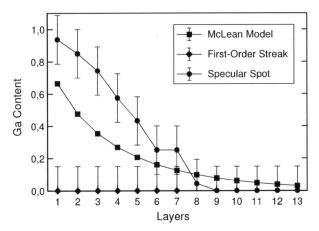

Fig. 10.2. Segregation profile determined with the relation of Fig. 10.1 at the specular spot position (*circles*) and the first-order streak (*diamonds*), and theoretical curve using the parameters of [276] (*squares*)

In any case, the ratio of surface to bulk composition is larger or equal for the calibration measurement. Since the RIPS is sensitive only to the surface composition, this means that the segregation profile of Fig. 10.2 establishes a *lower bound* for the true segregation as determined from the specular spot signal.

We can compare our results to theoretical models of segregation. The square data points in Fig. 10.2 were obtained from the thermodynamic mass action law on the basis of parameters obtained from independent experiments [276]. If there is an energy gain for an atom in a surface position compared to a bulk location, the chemical potentials of the bulk and surface are not equal for identical compositions. The surface and bulk concentrations of the elements deviate to align the chemical potential. This can be modeled using the entropy terms and a chemical-energy contribution E_s to the free energy. The balance of chemical potentials then yields McLean's equation [277]

$$\ln\left(\frac{x_b}{1-x_b}\right) + \frac{E_s}{kT} = \ln\left(\frac{x_s}{1-x_s}\right). \tag{10.1}$$

For positive E_s, the surface layer is enriched with the corresponding element; for negative energy it exhibits a deficiency of this element. The concentrations at the surface and in the bulk are denoted by x_s and x_b, respectively. This theoretical model has been checked to hold in many systems, at least for low x_b [278, 279].

If we assume that the third layer, counting from the surface, does not contribute to segregation, the equilibrium is established between the two top layers with the constraint of mass conservation. For the next deposited layer, the formerly second layer is fixed and the formerly first layer equilibrates with the new top layer. Values of 0.1 eV [276] and 850 K then yield the data shown by squares in Fig. 10.2. In the case of a heterointerface, the segregation profile depends heavily on the first layer of the new species, in which all of the segregating material available for the subsequent concentration gradient is removed from the substrate layer. For this first layer, x_b is initially unity. This indicates a limited validity of this model to describe heterointerface segregation, since it works best around concentrations of 0.5. However, the segregation profile of Fig. 10.2 has been found to agree with Raman data for short-period superlattices and is commonly used for comparison with experiments involving GaAs/AlAs heterointerfaces [27, 272, 276, 280].

The measured data imply a much stronger segregation and a convex instead of concave composition profile. This would raise doubts about the validity of our model if it actually represented the volume concentration gradient. On the higher-order streaks and in perpendicular azimuths, however, we do not see significant segregation; see, for example, Fig. 7.10. Using the model developed in Sect. 7.2.4, we can therefore associate segregation with the disordered regions of the surface, which means that segregation takes place preferentially at the step edges. This is what we expect, since the binding energy is smallest at steps and edges, where the atoms occupy sites with low coordination. The exchange process necessary for segregation is therefore most easily accomplished at step sites. On terraces, the energy barrier for a vertical exchange process is much higher since the terrace atom must be removed from its tightly bound site. Since the RIPS signal shows considerable segregation only at one spot along one azimuth, we can safely assume that segregation in this materials system is a localized process. A quantification, however, is difficult since there is no direct relation between the relative intensity of a particular reflection and the relative number of scatterers contributing to it. We are therefore limited to rather qualitative interpretations of our data.

In the literature [27,51,272,276,280–284], the models developed to describe segregation are one-dimensional in that they average over the lateral structure of the interface. This in turn means that they describe only processes that take place along the growth direction, perpendicular to the surface. A segregation mechanism at a step, however, is at least two-dimensional, depending on the symmetry of the step or edge site. In addition, some experiments [19,285] also average over interface roughness that is due to the undulation of the surface prior to interface formation. This roughness is not due to segregation, which additionally complicates the comparison with segregation models.

High real-space resolution can be achieved by HRTEM, with the restriction of averaging over the specimen thickness along the beam direction. As already discussed in Sect. 1.2, this specimen thickness is about 50 to 100 atomic planes. Two images, along the [$\bar{1}10$] and [110] azimuths, are shown in Fig. 10.3. Both images were taken with similar diffraction conditions to

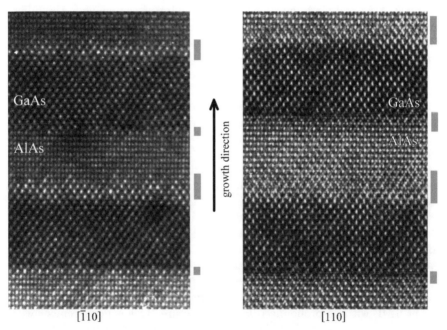

Fig. 10.3. HRTEM images along two azimuths of a nominal 14 ML/14 ML GaAs/AlAs superlattice. The electron energy was 400 keV. Both specimens were prepared from the same sample. The growth direction is upwards. The intermixed regions at both types of interface are indicated by the *gray bars* at the *right* of each panel

achieve comparable material contrast. Ideally, the unit cell of AlAs features two bright spots, whereas the GaAs unit cell shows only one bright area per cell.

Both HRTEM samples were taken from the same MBE-grown crystal in order to exhibit the same layer structure, under identical growth conditions, although at two different positions a few millimeters apart. The slightly different growth rates at the two positions produce different layer thicknesses in the two images. With the growth direction pointing upwards, one can clearly see the difference between the two types of interface. Whereas the inverted interfaces (GaAs grown on AlAs) are relatively abrupt, the normal interfaces (AlAs grown on GaAs) are mixed over several crystal planes. The intermixed areas are indicated by the gray bars to the right of each panel.

Since the spot intensities along the [110] and [$\bar{1}$10] axes generally depend in a complicated and nonmonotonic way on composition, a quantitative treatment is difficult [30]. We therefore restrict our discussion to a qualitative assessment. The thickness and defocus values, which determine the intensity distribution within the unit cells of the image, are reasonably constant across the area depicted. This can be concluded from the similar appearance of the different AlAs and GaAs layers away from the interface regions in the different layers. AlAs shows two spots per unit cell with equal intensity, while the GaAs contains only one bright spot per unit cell. Therefore, any deviations from these patterns indicate a material different from AlAs or GaAs. We therefore identify regions with unit cell contrast different from the GaAs or AlAs pattern as intermixed areas.

Along the [$\bar{1}$10] azimuth, the intermixed region is very narrow at the inverted interface, whereas the normal interface shows a distinct sawtooth structure distributed over four to five crystal planes along the growth direction. The strong lateral contrast indicates an arrangement with significant ordering along [$\bar{1}$10]. In view of the averaging over several tens of layers in the beam direction, the features that produce such good contrast must be extended along the viewing direction. At the same time, the lateral structure must be maintained over significant distances along the beam direction. We therefore conclude that the intermixed volume at the normal interface forms elongated mounds along [$\bar{1}$10] with a roughly triangular cross-section. At the inverted interface, the structure of the AlAs surface prior to interface formation is conserved, with an intermixing of at most two crystal planes.

In the [110] azimuth, the normal interface is intermixed over a range of about 6 ML with no pronounced lateral structure. This is in agreement with our interpretation of the [$\bar{1}$10] image. The elongated segregation volume does not show short-period variations along [$\bar{1}$10]. Its projection perpendicular to the elongation direction therefore appears uniform along the interface. Both interfaces in the [110] viewing direction are less sharply defined than in the [$\bar{1}$10] direction. This can be explained by the surface morphology of the initial GaAs surface, which serves as the template for interface formation.

Large anisotropy with long and relatively straight steps along [$\bar{1}$10] is usually observed; see Fig. 7.15a. The step density along [110] is therefore significantly higher than that required by the vicinality of the sample. The averaging in HRTEM includes more steps, resulting in a broadening of the interface that is due to the variations within the thickness of the sample. If we could prepare thinner samples, the laterally averaged interface width would most probably approach the values along [$\bar{1}$10].

How can we interpret this segregation behavior? In agreement with literature values [272], the saturation distance of the RIPS on the specular spot is almost independent of the growth rate. In contrast to the behavior expected on the assumption of kinetic limitation, it even increases with higher growth rate, as marked by the arrows in Fig. 7.9. Whereas the gray arrows in the figure indicate the transition distance of the $1.27\,\text{\AA}\,\text{s}^{-1}$ curve, the actual values marked by the black arrows become increasingly larger for higher growth rates. This suggests that the segregation process is close to thermodynamic equilibrium. The increase with growth rate can be explained by the fact that the surface roughness increases for higher growth rates, which increases the number of possible exchange sites for a step-mediated process.

The saturation distance of the RIPS on the specular spot is also a function of sample temperature; see Fig. 7.5. The gray arrows in Fig. 7.5 show an Arrhenius fit with an activation energy of $0.5\,\text{eV}$, compared to the transition distances determined from the experiment marked in black. Although the experimental data are noisy, the increase in transition distance with temperature would indicate a kinetic limitation, since the McLean model would predict a decrease of the segregation length with temperature.

The inconsistent behavior of the two measurements indicates the limited validity of a simple one-dimensional model to describe the segregation process. Instead, processes both parallel and perpendicular to the surface have to be taken into account. The surface dynamics during crystal growth involve nonequilibrium processes, as is evident from the presence of RHEED oscillations. There are fast and slow processes on the surface as we have seen in the analysis of the oscillation envelopes and recovery (Sect. 3.1). The movement of atoms on a terrace takes place on a much faster timescale than the movement of step-edge atoms or complete step systems. At the same time, an exchange process at a step where a molecule of material A is replaced by a molecule of material B is much easier at a step than on a terrace, where more bonds need to be broken for the exchange to take place. These different systems, each of them confined locally, can equilibrate on different timescales for different processes such as exchange reactions or surface migration, complicating the overall behavior of the segregation. It is probably only on this localized scale that we can talk of an equilibrium situation in MBE, since the equilibration of large surface areas, especially for sample orientations close to a singular plane, is clearly kinetically limited.

How can we explain the symmetric shape of the intermixed regions in Fig. 10.3, given that the step edges can move laterally during growth? A closer investigation reveals that during layer-by-layer growth, represented by the oscillating component of the RHEED signal, a previously present step on the surface is conserved. During homoepitaxy, the lower layer is filled, while an additional layer is added on the uphill side of the step and the step moves up one lattice plane. If we consider this layer formation to be fast compared to step and island propagation, such a step propagates almost perpendicular to the surface. These conditions are fulfilled for weakly damped oscillations.

For heteroepitaxy with possible segregation exchange at a step, several processes are possible. The situation for our hypothesis of a step-mediated process is shown in Fig. 10.4. In this simple atomistic picture, GaAs units

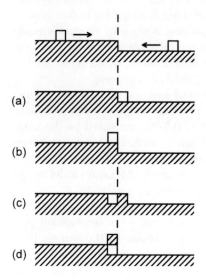

(a)

(b)

(c)

(d)

Fig. 10.4. Atomistic model for segregation at a heteroepitaxial step edge. GaAs units are denoted by *hatched areas* and AlAs units by *open squares*. For a GaAs unit approaching from the lower or upper level, any of the four final states (**a**)–(**d**) is possible, resulting in eight different events

are represented by hatched areas and AlAs units by open squares. For the two initial states of an AlAs unit approaching the step from above or below, the four final states directly after interaction with the step are shown in Figs. 10.4a–d; the corresponding processes are denoted 'a' to 'd'. Processes 'a' and 'b' represent attachment to the step, including a change of level for 'a' with approach from the left or 'b' with approach from the right. Processes 'c' and 'd' additionally involve the exchange of GaAs and AlAs units. This results in eight possibilities, four of which ('c' and 'd') lead to alloy formation at the step. A rate-equation model using these eight processes would involve eight transition probabilities, all of which are in principle unknown. Therefore, even a good fit to the experimental data would probably be of limited value. We can, however, immediately conclude that for a step-mediated segregation mechanism at the heterointerface, at least the 'd' processes must have a significant probability, since they are the ones providing GaAs material transport to the upper layer. The 'c' processes are responsible for lateral alloy formation. Both'c' and 'd' must be present to obtain the triangularly shaped segregation features observed in Fig. 10.4. Note that both processes 'c' and 'd' imply that *the center of the alloy* either moves slightly to the left or is stationary, in contrast to *the morphological step edge* position, which moves to the right with 'a' or 'c'. For equal deposition rates on the lower and upper terraces, the movement of the morphological step edge position depends on the net difference between the processes that move a unit up and down the step. This indicates that a step-mediated segregation process is able to produce the features observed in the Fig. 10.4 HRTEM cross sections, namely, laterally symmetric and well-aligned features, even if the morphological step edge moves because of a dominant step-down process. Instead, the important requirement is that the lateral center of the alloyed region remains stationary.

We can then propose the following model. As Al is deposited on the GaAs surface, AlAs units attach at the steps of the GaAs surface, and both in-plane exchange and pop-up processes affect the Ga atoms. From the asymmetry of the segregation at the two interfaces we can conclude that it is energetically favorable to replace an Al atom in AlAs by Ga, but that the replacement of a Ga atom in GaAs by Al costs some energy. The lateral exchange process will therefore come to a halt at a certain Ga/Al ratio. This ratio will depend on the concentration of the underlying layer, since the bonds to the substrate also count in the energy balance. The Ga that has been moved up one layer will stay close to the step edge since the upper layer is completed near the step and another equilibrium concentration is established there. This second-layer equilibrium concentration can be different from the first-layer value since the underlying layer is different now. The scheme can then be repeated until the Ga is used up or a certain residual surface concentration is maintained that is no longer incorporated into the growing crystal. If the localization of these mechanisms at the step is strong, it will lead to the observed laterally modulated segregation profile.

Two unresolved points remain for further study. First, we would like to determine segregation quantitatively on an atomic scale. This could be done by STM, or, if atomic resolution with chemical contrast is possible, by atomic force microscopy (AFM) of cleaved samples. These experiments allow the identification of atoms at the cleaved surface and therefore the determination of the composition of a single atomic plane in the planes of both images shown in Fig. 10.3. Second, a possible equilibrium composition would allow the growth of sharply defined $Al_xGa_{1-x}As$/GaAs heterostructures with $x = x_{Equilibrium}$. RIPS measurements could be used to determine $x_{Equilibrium}$. The very low damping of RHEED oscillations for $x = 0.5$ (Fig. 3.4) suggests that $x_{Equilibrium}$ is close to 0.5.

Summarizing the RHEED, STM and HRTEM results, we obtain the following interface structures of GaAs and AlAs. At the normal AlAs-on-GaAs interface, mounds of $Al_xGa_{1-x}As$ form with a height of several monolayers and a similar width along [110]. They are strongly elongated along [$\bar{1}$10]. At the inverted interface, the morphology of the AlAs surface is conserved, with intermixing below the detection level of the methods used. From Fig. 7.16 we can conclude that the AlAs surface is less anisotropic than the GaAs surface. This would lead to different anisotropies at the two interfaces of a quantum well. Whether this surface anisotropy is related to the surface reconstruction, and could therefore be controlled by the growth parameters, is not clear at this time.

Comparing our results to the literature data presented in Sect. 1.2, we can confirm the segregation at the normal interface. The assumption of laterally uniform segregation generally used, however, has to be modified to incorporate an anisotropically modulated composition profile. The input of our model of the interface structure into theoretical calculations could show whether a better fit for optical data, especially the peak splitting in photoluminescence, can be achieved.

10.1.2 (113)A Interfaces

In this section, we investigate the (113)A surface of the GaAs/AlAs materials system. If we can perform a RIPS measurement on this surface orientation, we should be able to characterize the interface structure similarly to the (001) case. With an abrupt interface, the corrugation of the surface reconstruction should be preserved at the interface. The RHEED patterns are very similar for both the GaAs and the AlAs surface, as can be verified from Fig. 10.8. This means that from a structural point of view, both interfaces can be expected to be similar since the surface reconstructions are the same. On the other hand, this close similarity implies that we cannot expect any RIPS. Nevertheless, some conclusions can be drawn from the RHEED dynamics.

Figure 10.5 shows the pulse sequence we used on the (113)A surface. Since the crystal plane spacing is only 1.7 Å and the oscillations are strongly damped, the growth pulses were shortened in favor of longer recovery intervals, which improve the relative smoothness of the surface prior to growth. The pulse sequence again consisted of two GaAs intervals, followed by three AlAs and two GaAs pulses. RHEED oscillations could only be observed at very low incidence angles on the central streak close to the intensity minimum of the reconstruction. This location corresponds to the area just above the left-hand end of the length marker on the (00) streak in Fig. 4.8. It represents a kinematically forbidden reflection, again confirming the multiple-scattering origin of RHEED intensity oscillations. The geometry is shown in the inset of Fig. 10.5. Figures 10.6 and 10.7 show the temperature dependence of the second GaAs interval and the first AlAs pulse. All other growth and diffraction conditions are identical to those of Fig. 10.4. The GaAs oscillations in Fig. 10.6 show a pronounced amplitude maximum around 600 °C. This coincides with the brightest and most clearly defined RHEED pattern.

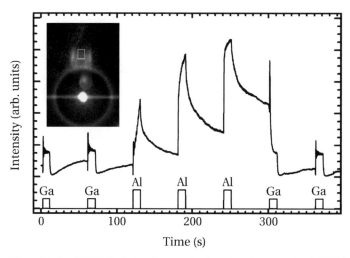

Fig. 10.5. RHEED intensity evolution for the standard RIPS measurement sequence on a (113)A surface along the $[\bar{1}10]$ azimuth. The measurement geometry is shown in the inset. The beam energy was 20 keV, with an incidence angle of $0.7°$. The sample temperature was $602\,°C$, using an As_4 pressure of 1.7×10^{-3} Pa. The growth rates were $1.5\,\text{Å s}^{-1}$ for GaAs and $1.3\,\text{Å s}^{-1}$ for AlAs

At temperatures above $\approx 610\,°C$, desorption sets in, which becomes visible as an increase of the oscillation period.

For AlAs, oscillations are observed only at the normal heterointerface close to the GaAs. The temperature dependence of these oscillations is shown in Fig. 10.7. As for GaAs, the most pronounced oscillations are found around $600\,°C$. The absence of oscillations in the AlAs part of the sequence indicates that these oscillations are due to the GaAs, either because of a smoother GaAs surface prior to AlAs growth or because of segregation. Since the RHEED patterns of both surfaces are very similar (Fig. 10.8), we do not expect, however, that the surface morphologies of the surfaces are drastically different.

The absence of oscillations inhibits the evaluation of the oscillation phase. We can see, however, that the intensity evolution during normal heterointerface formation shows some peculiar features. During growth, there is a strong increase of the average intensity in the measurement window. This could be directly attributed to the growth dynamics, but in the next growth pulse, the average intensity continues at the same value as at the end of the preceding growth interval. It is therefore, obviously, not influenced by the rearrangement of the surface atoms during growth interruption. This means that the average RHEED intensity at this position of the RHEED pattern monitors some property of the top crystal layer other than its ordering. We therefore think that it directly represents the alloy composition at the surface, since the reconstruction is similar on both sides of the heterointerface. In this case, the difference in atomic form factors can change the intensities in the pattern

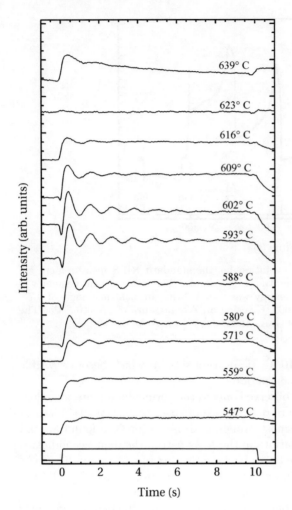

Fig. 10.6. Temperature dependence of the oscillations from the second GaAs growth pulse of the sequence. The amplitude of the oscillations peaks at 600 °C. The growth and diffraction conditions are identical to those of Fig. 10.5

similarly to the discussion in Sect. 7.2.2, leading to a correspondence of the intensity of certain reflections with the material composition.

Again, the behavior of this average intensity is asymmetric, similar to the RIPS on (001). After a slow increase at the normal heterointerface, the intensity drops abruptly to the previous GaAs level at the inverted interface; see Fig. 10.5. We can therefore conclude that significant long-range segregation is present at the normal heterointerface, while the inverted interface is abrupt. This is in agreement with our results described in Sect. 10.2, where segregation is also found to be independent of surface reconstruction. As for Sn, the effect depends only weakly on the temperature. This can already be seen from an inspection of Fig. 10.7, in which all traces show the same slope. An overview of the AlAs pulses for three temperatures across the range probed is given in Fig. 10.9. The total increase of intensity during growth, as well

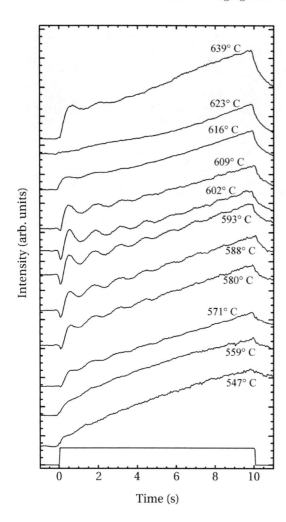

Fig. 10.7. RHEED oscillations versus temperature for the first AlAs growth pulse of the measurement sequence of Fig. 10.5. The oscillation amplitude shows a maximum at the same position as for GaAs homoepitaxy. The average slope of the curves is very similar

as the individual pulse shapes, is very similar for all three curves. This weak dependence on temperature is in agreement with our assumption of a (local) equilibration process similar to a 2D McLean model.

We therefore conclude that segregation of Ga at the normal heterointerface is a process independent of crystal orientation and surface reconstruction. The third AlAs pulse, after the deposition of 40 Å, still shows a significant slope. If our interpretation in terms of alloy composition is correct, we still have significant segregation this far from the normal interface. In the absence of calibration measurements on this surface, the longer decay time of the signal on (113)A suggests an enhanced segregation on this surface compared to (001). This is in agreement with studies for In segregation that find enhanced segregation on $(n11)$ surfaces for small n [286].

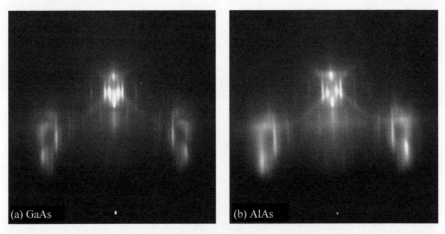

Fig. 10.8. RHEED patterns for the (**a**) GaAs and (**b**) AlAs (113)A crystal surfaces along [$\bar{1}$10]. The growth and diffraction conditions for both patterns are identical to those of Fig. 5.11, which corresponds to the same misorientation

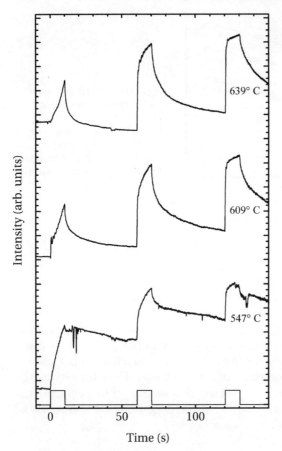

Fig. 10.9. Variation of the three AlAs growth intervals with substrate temperature. The growth and diffraction conditions are identical to those of Fig. 10.5

The corrugation of the surface reconstruction can therefore be expected to be present at the inverted heterointerface of (113)A, whereas the interface structure at the normal interface should be significantly altered, if not dominated, by segregation. This is confirmed by Raman spectroscopy measurements on these structures [200]. In these measurements, the shifts of the GaAs-like confined optical phonons are used to determine the width of the intermixed region. At the same time, the splitting of the confined LO_3 modes confirms the presence of an interface corrugation. Calibration measurements of the RHEED intensity during growth as a function of substrate composition could provide a quantitative evaluation method.

10.2 Modifying the Surface Reconstruction: Tin Doping

We have seen in the previous section that the two different GaAs/AlAs heterointerfaces differ significantly in structure and composition profile. The results suggest that segregation takes place at the surface steps and therefore depends on the step structure. Since segregation is a surface process, the question arises as to whether or not it also depends on the surface reconstruction.

An experiment of this type can be realized by depositing Sn on the (001) crystal surface. Sn is known to strongly segregate on GaAs and AlAs with a surface-layer-to-bulk concentration ratio of > 1000 for standard growth conditions [287, 288]. On the one hand, this complicates a reliable determination of Sn surface concentrations under MBE conditions, which was therefore not done for the experiments described in this section. On the other hand, it allows us to predeposit a certain amount of Sn on the crystal surface and then perform the usual pulsed sequence for GaAs and AlAs. Since very little of the segregating layer is incorporated during growth, the surface concentration remains virtually unchanged during the measurement. This allows us to perform the RIPS sequence with conditions identical to the standard GaAs/AlAs case, but with a different set of surface reconstructions.

Figure 10.10 shows a part of such a growth sequence. The substrate temperature was 570 °C and the As_4 pressure was adjusted so that the AlAs/GaAs growth sequence was in the prehump region with monotonic phase shift, similarly to the experiments in the last section. The electron energy was 20 keV at an incidence angle of 1.2 degrees. With increasing Sn coverage, a surface reconstruction with threefold superstructure along $[\bar{1}10]$ emerges. Whereas the reconstruction is symmetric for GaAs, it changes to an asymmetric structure for AlAs. The gradual transition can clearly be observed in the first AlAs growth interval. For still higher coverages, both GaAs and AlAs show the asymmetric threefold structure. We can safely assume that the surface coverage of Sn is a significant fraction of an ML, since a number of Sn atoms on the order of the number of the surface unit cells can be assumed necessary to switch the surface reconstruction. On the other hand, we do not

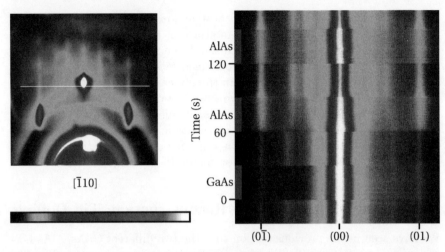

Fig. 10.10. Part of the standard RIPS growth sequence of GaAs/AlAs with segregating Sn producing different surface reconstructions. The position of the measurement line and the color conversion are indicated on the *left*. See color plates at the end of the book

expect the coverage to exceed 1 ML, since Sn then starts forming droplets that cause a strong increase in the RHEED background signal. The surface reconstruction at the end of the growth sequence was not significantly different from the reconstruction at the beginning, confirming our assumption of a small incorporation rate. The oscillations of the specular spot of Fig. 10.10 (the central part of the line shown) are given in Fig. 10.11 together with the signal of the first-order reflection on the Laue circle.

The RIPS is different from that in the Sn-free case. This is expected, since both surface reconstructions have changed and therefore, presumably, also their relative group III element contents. However, the phase shift takes place immediately on the specular spot. At first sight, we might interpret this as an absence of segregation in this case. The slowly adjusting envelope shape, however, points to a situation similar to that for the [110] azimuth on the unmodified surface, which is also not very sensitive to segregation (see Fig. 7.10). This would mean that the surface is less anisotropic or that the anisotropy direction is perpendicular to that of the unmodified GaAs surface. This assumption is also supported by the gradually changing fractional-order streak separation at the normal interface. Looking at the first AlAs growth interval of Fig. 10.10, one can clearly see that the peak separation does not change abruptly at the interface, but adjusts gradually during several monolayers, indicating that Ga segregation takes place at the Sn-modified surface as well.

Using our knowledge of the top-layer interference model discussed in Chap. 9, we can visualize this segregation more directly. Obviously, it contributes only very little to the RHEED oscillation signal along this azimuth

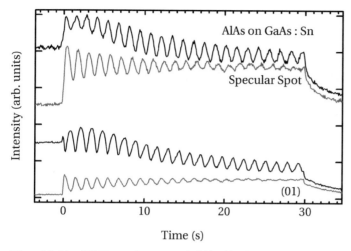

Fig. 10.11. RIPS at the normal AlAs/GaAs interface for a surface covered by a fractional ML of Sn. The data are measured at the specular spot position of Fig. 10.10

and is not very sensitive to changes in the surface reconstruction. The sensitivity, however, can be enormously increased by adjusting the incidence angle to a position very close to the generalized Bragg conditions of the phase rocking curve, similarly to the simulations of Fig. 9.13. This is shown in Fig. 10.12. The sequence was grown under growth conditions that produce the transitional hump phase at the normal heterointerface for undoped GaAs. The temperature was the same as in Fig. 10.11 to avoid effects due to temperature variation. The Sn is seen to suppress the hump, in agreement with our conclusion that the hump is caused by a surface reconstruction change. As can be verified from Fig. 10.10, the surface reconstruction changes gradually and continuously at the interface, without discontinuities.

The transitional range of additional maxima at the normal interface clearly demonstrates the gradual change of surface composition also seen in the reconstruction evolution. The same doubled extrema are seen at the inverted interface, but again the transition at this interface is accommodated during the first ML of growth. This means that the segregation behavior is similar to the pure GaAs/AlAs case, with very similar transition distances. We therefore conclude that the segregation mechanism is not significantly modified by the different surface reconstruction. This does not mean that the segregation is quantitatively identical to that in the nonmodified surface, since we have to consider that it may depend on the step density prior to interface formation and possibly other factors as well.

The identical transition distances for both surfaces, however, indicate that the segregation mechanism is independent of the surface reconstruction. This is in agreement with our assumption of a fast process that approaches thermal

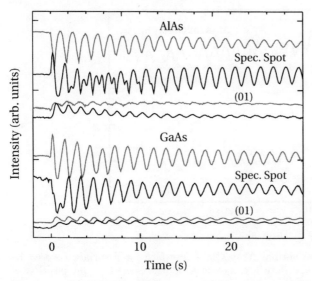

Fig. 10.12. RHEED oscillation signal from the normal and inverted GaAs/AlAs interfaces of the Sn-covered system. The growth conditions were similar to those of Figs. 10.10 and 10.11, except that the As_4 pressure was lower so that the measurement without Sn features the familiar hump about 5 ML wide

equilibrium in the neighborhood of the steps. The exchange energy driving the segregation seems to be a property of the material pair GaAs/AlAs, largely independent of their surface structures and, if we consider the results of the previous section, also independent of the crystal orientation. We can therefore expect similar compositional profiles in all cases.

10.3 Modifying the Surface Morphology: Carbon Doping

In this section, we study the effect of carbon δ-doping on the phase of the oscillations during overgrowth. Carbon doping is known to reduce the lattice constant of GaAs [289–291], thereby introducing strain at the heterointerface that modifies the overgrowth behavior. It is also generally incorporated as a shallow acceptor that is ionized at the growth temperature. The intrinsic carrier density of pure GaAs is quite low; at a typical growth temperature of $580\,^{\circ}\mathrm{C}$ it amounts to approximately $2\times10^{16}\,\mathrm{cm}^{-3}$ [292]. This means that, in addition to strain effects, the dopant type and concentration become important since the carriers introduced by doping are not masked by the intrinsic carrier densities.

The results of STM experiments on doped GaAs surfaces can be explained by the following model [293]. At an undoped $\beta(2\times4)$ reconstructed surface, the Fermi level is pinned at midgap because of intrinsic surface defects such

as missing unit cells or steps. For n-type doping, kinks in the As dimer rows
of the reconstruction form that act as surface acceptors and keep the Fermi
level at midgap, even for very high dopant concentrations. For p-type doping,
no donor-like surface state exists, and after the intrinsic defects have been
compensated, the Fermi level approaches the valence band. This means that
for p-type doping, the $\beta(2\times4)$ surface remains unchanged or becomes even
more ordered. This can also be observed in RHEED. Figure 10.13 features
two profiles across the reconstruction in the $[\bar{1}10]$ azimuth as a function of

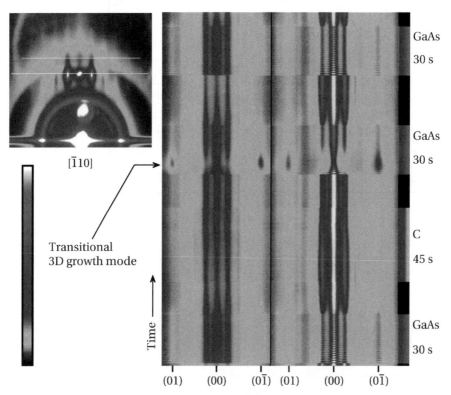

Fig. 10.13. Linescans of a growth sequence with a δ-doping C layer in GaAs.
The line positions in the pattern and the color conversion are shown to the *left*.
During the first overgrowth the surface roughens, which can be deduced from the
transient appearance of transmission spots in the RHEED pattern. The As$_4$ BEP
was 4.7×10^{-3} Pa, with a sample temperature of $570\,^\circ$C. See color plates at the end
of the book

time. From the constant spacing of the fractional-order streaks we can deduce
that the surface reconstruction does not change during C deposition, even at
the high 2D density of 9.5×10^{13} cm^{-2} used in the experiment. This means
that there cannot be any resulting phase shift at the overgrowth interface,
since the group III element content at the interface does not change. There

should also be no diffraction-induced shifts, since the unit cell structure remains unchanged. During the first overgrowth at high predeposited carbon concentrations, we observe a transient roughening of the growth front. This can be concluded from the appearance of bulk diffraction spots and the strong decrease in the amplitude of the oscillations.

A plot of the RHEED oscillations of Fig. 10.13 is shown in Fig. 10.14. The dark-gray curve shows the GaAs-on-GaAs reference interval prior to C

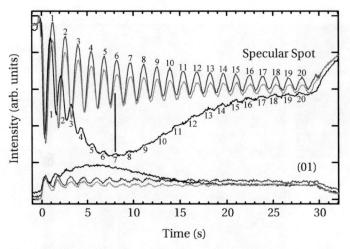

Fig. 10.14. Oscillations recorded at the specular spot position and on the first-order streak of the experiment of Fig. 10.13. The roughening of the growth front is accompanied by a strong phase shift of more than one period on the specular spot

deposition; black denotes the first and light gray the second overgrowth pulse. During the first overgrowth, the phase shifts strongly on the specular and the first-order spot, reaching a value of more than one period around the seventh oscillation. This position is marked by a vertical line and corresponds to the minimum of the intensity of the specular spot. At the same time, the black oscillations are in registry with the reference at the beginning as well as the end of the oscillation interval. This confirms our prediction from the behavior of the reconstruction streaks that neither diffraction-induced shifts nor shifts due to different group III element contents are expected. The observed phase shifts must, therefore, be due to the morphological changes of the growth front at the heterointerface.

We can find an explanation of these morphologically induced phase shifts by taking into account the small sampling depth of the RHEED electrons. For the conditions of our experiment, we are approaching conditions where the strain is sufficient to cause undulations in the surface. For the surface to become unstable in such a way, the growth rate at the top surfaces of the forming mounds must be higher than in the valleys in between. If the average slope becomes larger than the incidence and exit angles of the RHEED electrons, the valleys are no longer visible to the incoming electrons. The RHEED oscillations then show only the higher growth rate on the top of the mounds, and the phase position shifts. At the same time, a transmission pattern is generated from the peak regions with a thickness in the beam direction smaller than the extinction distance. For the largest modulation of the surface, the most intense transmission pattern as well as the largest phase shift is expected. At this point, marked by the vertical line in Fig. 10.14, the growth rate on the peaks deviates the most from the equilibrium value and the peaks are at their narrowest. After the maximum of the surface modulation, the valleys fill in again and the growth rate again approaches a constant value everywhere. For this to take place, the growth rate at the peaks must become smaller than the average value, which manifests itself in a phase shift in the opposite direction. After the surface has recovered to smooth layer-by-layer growth, the oscillations are in phase with the reference case of continuous smooth growth, since the same amount of material is deposited in both cases and the surface reconstruction has not changed.

The region of disturbed growth is located close to the interface. The second overgrowth interval is is no longer affected by the tensile stress of the carbon. This indicates that C does not segregate over large distances. If it did, the second overgrowth interval would resemble the first. Instead, the second overgrowth pulse is practically identical to the reference.

An explanation of the beating observed on Sn-doped surfaces [163] has been proposed that attributes the beating to a similar undulation of the surface during growth [164]; see Sect. 3.1.

10.4 Silicon Doping

Having studied the effects of C δ-doping, we turn to Si. Si is predominantly n-type in (001)-grown GaAs, and it segregates. Both properties significantly change the signal we obtain in a RIPS experiment as well as the information we can extract from such a measurement.

10.4.1 Dependence of the Phase Shift on Si Concentration

An extension of our standard RIPS measurement sequence to characterize Si δ-doping is shown in Fig. 10.15. Si is deposited on both the GaAs and AlAs surfaces in the δ-doping mode at a rate of $3.8 \times 10^{11}\,\mathrm{cm}^{-2}\,^{-1}$. This rate was determined by secondary-ion mass spectroscopy (SIMS) of a volume-doped reference sample.

The separation between the overgrowth intervals was kept constant so that the surface configuration for both growths was as similar as possible. The Si deposition time was varied from 5 s to 100 s in a series of experiments

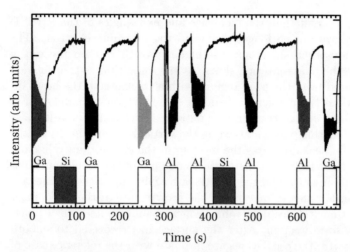

Fig. 10.15. RIPS measurement sequence for the silicon-doping experiments. The substrate size was 4×4 mm. The substrate temperature was 570 °C, with an As_4 BEP of 2.3×10^{-3} Pa. The growth rates of both GaAs and AlAs were $2 \, \text{Å s}^{-1}$. The electron beam incidence angle was 0.85°

otherwise identical to that shown in Fig. 10.15. For this range of Si coverages, the reconstruction transition to the threefold periodicity [209] was not reached. The RHEED intensities were recorded on the specular spot as well as at the intersection of the Laue circle with the 01 rods along the [$\bar{1}$10] azimuth.

For GaAs growth, the resulting RHEED intensity oscillations of the growth interval preceding Si deposition, as well as of the two intervals following the Si deposition, are shown in Fig. 10.16. The upper three traces show the intensity variation on the (00) streak, whereas the lower three curves were simultaneously recorded on the (01) reflection. For the first overgrowth interval (black line), the oscillations are strongly damped and shifted to the right by almost half a period. The second growth interval (light-gray curve), starting 21 crystal planes above the nominal Si plane, still shows a significant phase shift of the oscillations on both the (00) and the (01) streak. When we plot the phase shift of the first few periods of the RHEED intensity oscillations after Si deposition as a function of the amount of Si deposited, we obtain the relationship shown in Fig. 10.17. As indicated by the least-squares fit to the data, the dependence of the phase shift on Si concentration is approximately linear.

The rate of the shift amounts to $0.13 \times 10^{-13} \, \text{cm}^2 \, \text{atoms}^{-1}$. The deviations observed for 1.3 and $2 \times 10^{13} \, \text{cm}^{-2}$ can be attributed to segregation effects as discussed in the next section. The first six experiments were made back to back, whereas for the highest doping level a thick buffer layer of GaAs was predeposited to bury any residual Si and separate any free carriers from the surface layer. The observed linearity of the phase shift is remarkable, consid-

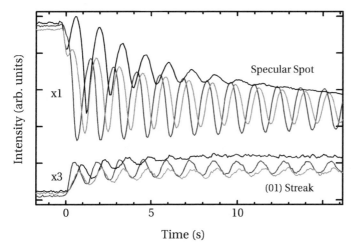

Fig. 10.16. Oscillations for GaAs growth using the sequence of Fig. 10.15 with 100 s Si deposition time, resulting in 3.8×10^{13} cm^{-2} Si. The curves from the (01) streaks are expanded by a factor of 3. *Dark gray* denotes the unmodified reference, and *black* the first and *light gray* the second overgrowth after Si deposition

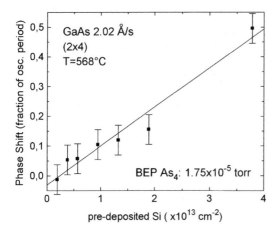

Fig. 10.17. Phase position of the first RHEED oscillation during GaAs overgrowth as a function of predeposited Si. The shift shows an approximately linear dependence on Si surface concentration

ering the complicated dependence of the oscillation phase on the diffraction conditions (Fig. 3.5).

Using the model of Sect. 9.2, we can interpret the phase shift in terms of a change in the layer potential or in the thickness of the sampled layer, or both, using the top-layer interference model. For example, a change of the real or imaginary part of the potential as shown in Fig. 9.4 or 9.5 at an angle of 0.85° changes the phase by about one-third of a period over the range plotted. This is too small for the observed magnitude of the shift, but the simulations away from the phase jumps often show an almost linear behavior similar to that in our experiments. This suggests that the model of Sect. 9.2

can describe the basic mechanism for the observed phase changes due to the Si-induced modification of the surface.

10.4.2 Si-Induced Kinks

STM scans allow us to study the surface reconstruction changes induced by the deposition of Si in greater detail. Single Si donors have been detected by STM when the cleaved (110) plane of doped GaAs crystals was probed [33]. On the (001) as-grown surface, however, localized Si donors similar to the ones on the cleavage plane have not been reported. This is very likely due to a compensation mechanism at the reconstructed surface [294, 295]: kinks are introduced in the dimer–vacancy rows of the (2×4) surface reconstruction that accommodate the additional electron from the Si donor. This process is shown in Fig. 10.18. As expected from the simulation of Fig. 4.23, the half-

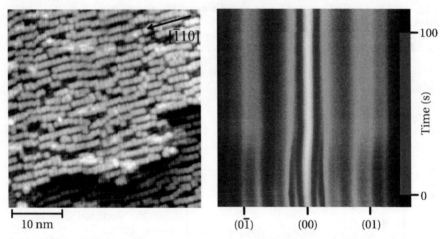

$[\bar{1}10]$ 10 nm $(0\bar{1})$ (00) (01) Time (s)   100   0

Fig. 10.18. STM image of a GaAs $\beta(2\times4)$ surface after the deposition of $3\times10^{13}\,\mathrm{cm}^{-2}$ Si, and corresponding intensity evolution of the reconstruction streaks. The 100 s deposition interval corresponds to a final coverage of $3.8\times10^{13}\,\mathrm{cm}^{-2}$. See color plates at the end of the book

order streak weakens and vanishes with increasing Si coverage and therefore increasing kink density. The number of kink sites is approximately proportional to the donor concentration [294].

We performed RIPS experiments at different sample temperatures and did not see significant phase shifts if the surface reconstruction was different from $\beta(2\times4)$; this was true for the adjacent $\alpha(2\times4)$ and $\gamma(2\times4)$ as well as the higher-temperature (3×1) and the lower-temperature $c(4\times4)$ reconstructions. This represents strong evidence that the mechanism causing the phase shift is closely related to the modification of the surface by the dimer-row kinks. It does not mean that the sample surface is not modified by Si in the cases where no shift is present. We just do not see any evidence of such changes in RHEED. The absence of phase shifts in these cases also means that there is no redistribution of the growth front into more monolayers for temperatures higher and lower than the $\beta(2\times4)$ range, similarly to the C doping case. The absence of such a morphology-induced shift allows us to assume that a growth-front roughening is also absent for the $\beta(2\times4)$ structure. Therefore the observed phase shift must be due to changes in the reconstruction only.

10.4.3 Si Segregation

When we attempt to measure the segregation behavior of Si on GaAs (001), the situation becomes more complicated. For bulk doping, the kink density on the surface has been shown to first increase during growth on a semiinsulating substrate. After a certain distance it then saturates at a constant value depending on the dopant concentration [294]. It is also known [296] that the carrier density of δ-doped samples saturates at about $1\times10^{13}\,\mathrm{cm}^{-2}$ and compensation then sets in. Segregation of Si has been investigated by several authors [296–302] using different measurement methods. Segregation distances of up to about 50 nm were found. To explain these observations, a model was proposed [297] in which the electric field introduced by Fermi-level pinning at the surface is considered as the main driving force for segregation. Strong band bending corresponding to the activation of all donors close to the surface was observed at doping levels that lead to compensation in the bulk material. This was explained by a site change of the dopant atoms when leaving the depletion region. Tunneling spectroscopy studies of the kinks in the surface reconstruction [294, 295] have revealed the presence of one negative charge per kink site, identifying these defects as acceptors. The number of kink sites was then found to equal the number of Si donors within the depletion region by analyzing the kink density saturation in volume-doped samples.

These findings confront us with the difficulty of correlating the phase of RHEED oscillations with the *surface* layer concentration of Si. If the integrated donor density of the whole depletion layer determines the kink density, the relationship between the dopant profile during overgrowth and the phase cannot be determined directly. Self-consistent band structure calculations would be needed to calculate the dopant distribution. Our type of measurement cannot directly show whether this long-range interaction between donor and kink exists or not, nor can we determine whether it corresponds to the depletion layer thickness, if present.

We do think, however, that the assumption of long-range interaction as the reason for compensation is not very probable. The postulated site change of Si when leaving the depletion layer involves the transport of a Ga atom from the surface to the Si site as well as the migration of the displaced As atom back to the sample surface. Ab initio calculations have found a lowest possible activation energy of 1.9 eV for a Si atom to leave a $\mathrm{Si_{Ga}}$ site. The migration barrier for interstitial Ga atoms was calculated to be of the same magnitude [303]. Electric field energies and thermal activation energies at the actual growth temperature are much smaller. In addition, Ga at defect sites is positively charged, except when substituting for As [304]. This means that it should migrate *against* the electric field direction established by the band bending. The energy necessary for this multiparticle process and the annihilation of the kink would have to be provided by the energy change of the charged donor due to band bending. We therefore conclude that the site change of the Si so as to be incorporated as an acceptor very likely takes place at the surface when the atom leaves the kink site.

This does not mean that the Si atoms are not able to diffuse. Diffusion is a process with a comparatively low activation energy [303], and concentration gradients are large for δ-doping. Diffusion has been observed in SIMS profiles of δ-doped layers, in addition to segregation [301]. Diffusion of Si from the surface into the bulk is facilitated by the fact that the process with the lowest activation energy is the migration of a negatively charged $(Si_{Ga}-V_{Ga})^{2-}$ complex, which means that it takes place in the direction of the field due to band bending.

The increase and saturation of the kink density for the growth of volume-doped GaAs can be explained by segregation only. Studies of the (110) cleavage plane of GaAs [34, 305] predict and show strongly localized features in the STM images of dopants in the top layer. These features are connected with dangling-bond states located near midgap. If we assume a similar behavior on (001), we would expect to see localized features also, if the surface were not reconstructed. In the reconstruction, on the other hand, the formation of a number of kinks equal to the number of deposited Si atoms is observed. We thus identify each surface Si atom with a kink site and combine this with segregation. When a Si atom arrives at the surface, it forms a kink site complex and gains some energy. To be incorporated into the crystal, this complex must dissociate again and the energy barrier needs to be overcome. This can be regarded as the reason for segregation. As more and more Si arrives, the density of kink complexes increases until, in any given time interval, the number of Si atoms incorporated equals the number of Si atoms arriving. Steric effects at the surface or p-type background doping can lower the energy barrier for incorporation and therefore reduce segregation. The compensation observed at high doping levels should take place at the growing surface when the Si atom is detached from the kink complex and incorporated into the bulk. The formation energies of Si_{Ga} and Si_{As} are comparable at high n-type doping [303] and therefore incorporation at high doping levels becomes amphoteric. The creation of kink sites, producing one negative charge per unit, is probably limited only by steric effects, and kink concentrations corresponding to the observed high activation of donors near the surface can easily be reached [306, 307].

If there is no long-range separation between the Si donor and its corresponding midgap state, our measurement method provides an in-situ and real-time method to directly measure dopant profiles. A RIPS measurement, however, does not directly provide information about the separation between donors and kinks. From the Bohr radius of a donor in GaAs or the depletion widths at high doping levels, this distance can be expected to be considerably larger than a few atomic distances.

The decay of the phase during overgrowth of the nominal δ Si layer by GaAs is depicted in Fig. 10.19. The data were taken on the specular spot for

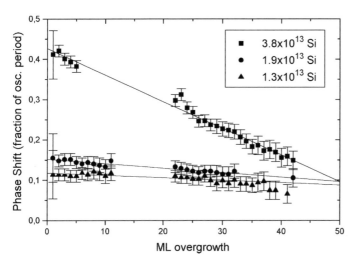

Fig. 10.19. Phase shift decay during overgrowth of GaAs for three different concentrations of predeposited Si

the three highest doping levels of Fig. 10.17 during the subsequent growth periods. The sequence of the measurements is from bottom to top, with a buffer layer grown before the last measurement was made. The first, larger error bar indicates the absolute uncertainty in phase determination and the smaller error bars show the relative error of the subsequent oscillations with respect to the position of the first point.

All three sets of data show a fairly linear decay of the phase shift. Since we calibrated the phase shift with surface instead of bulk concentrations, it does not directly correspond to the segregation profile as in the GaAs/AlAs case. Instead, the amount of incorporated Si in every layer is the difference between the phase shifts of two successive oscillations if we assume local interaction. We therefore obtain almost rectangular segregation profiles.

This agrees well with SIMS data [301, 302, 308]. For ultrahigh δ-doping concentrations approaching 1 ML, the typical profile consists of a broadened spike centered at the nominal layer position and a flat-topped shoulder with a Si concentration of approximately $5 \times 10^{18} \, \text{cm}^{-3}$, the typical threshold for compensation. If we combine this with the onset of compensation in the δ-doping case, we retain a surface concentration of roughly $2 \times 10^{13} \, \text{cm}^{-2}$ when compensation stops. This concentration distributed at a constant concentration of $5 \times 10^{18} \, \text{cm}^{-3}$ yields a layer thickness of approximately $300 \, \text{Å}$. This is in good agreement with the SIMS profiles. Applied to Fig. 10.19, this means that for surface concentrations above the threshold for compensation, additional Si atoms are incorporated at nondonor sites so that the surface kink concentration and therefore the phase shift decrease rapidly. As soon as the compensation threshold is reached, however, the dopant atoms are incorporated at a constant rate determined by the maximum activation level in the bulk. The incorporation rate is then much smaller, resulting in a smaller slope of the phase decay.

The projected intersection with zero of the linear fits for the two lower curves agrees well with our layer thickness estimate of about $300 \, \text{Å}$. The good agreement between SIMS and RIPS data, however, does not provide information about the interaction distance of donor atoms and kinks, since the depth resolution of SIMS is about $100 \, \text{Å}$, which is of the same order as the probable interaction distance.

On the other hand, we can be quite confident that, for the atoms incorporated at nondonor sites, we measure the profile with near-atomic resolution. The site change from donor in the surface position to nondonor in the bulk very likely happens upon detachment from the surface. This means that the corresponding kink vanishes at the same time, contributing to the phase shift. Therefore, the nonactivated Si incorporation rate is directly measured, whereas for the electrically active contribution we can only assert that it is below the $100 \, \text{Å}$ resolution of SIMS.

As in most other systems investigated, the phase shifts on the fundamental and first-order streaks are also different for Si δ-doping. A close-up of the first five oscillations of Fig. 10.16 is shown in Fig. 10.20. Again, the top three curves were recorded on the (00) streak and the bottom three curves on (01). The dependence of the phase on the Si concentration is also linear on the (01) reflection; see Fig. 10.21. The phase shifts significantly less on (01) in the second overgrowth interval. This means that the segregation of Si atoms is enhanced at the step edges. The difference between the first-order and specular spot signals, however, is not as strong as for GaAs/AlAs (Fig. 7.11). This means that segregation is not restricted to disordered areas. The mechanism of the process must therefore be different from that of Ga segregation in AlAs. A linear decay of the Si surface concentration means that the incorporation rate during growth is constant and does not depend on the amount of Si present at the surface. This points to an equilibrium model, but different from the one based on McLean's equation that we used for GaAs on AlAs. Instead, GaAs beneath a Si surface layer seems to readily incorporate Si atoms until the activation threshold is reached. This holds as long as the

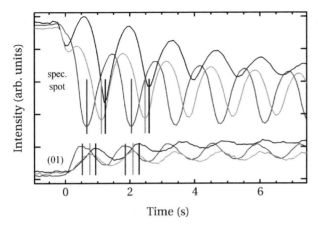

Fig. 10.20. Comparison of the phase shifts on the specular spot and (01) streak. The phase decays faster on the (01) streak

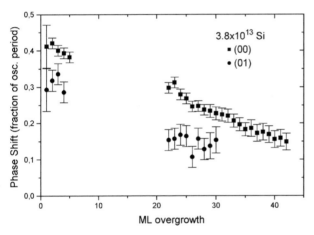

Fig. 10.21. Comparison of the phase shift decay during first and second GaAs overgrowths of the Si δ-doping layer. In our model, the slope of the lower curve represents the bulk segregation, whereas the top trace monitors the Si concentration at steps running along the beam direction

surface concentration is below 2×10^{13} cm^{-2}. The constant slope of the phase decay indicates that equilibrium with the bulk concentration is maintained during growth. For small surface concentrations, we expect a deviation from a constant incorporation rate as soon as the equilibrium cannot be reached with the time constants involved in the growth. This could be the case for the last few points of the lowest data set in Fig. 10.19. Above the compensation threshold, a similar mechanism seems to apply since the slope of the data here is also constant. Evaluation of the slope yields a bulk concentration of 1.7×10^{19} cm^{-3} in this range.

When the same experiment is performed with AlAs instead of GaAs, the phase shift of the overgrowth intervals is zero at the onset of the oscillations. During growth, the phase slowly shifts to about one period after 20 oscillations in the first overgrowth and half a period in the second. Since no reconstruction change upon Si deposition is observed, we assume that the behavior for AlAs growth is similar to the carbon-doping case. However, modified growth sequences with thick predeposited AlAs buffers need to be used to exclude Ga segregation effects and to see whether a saturation of the phase shift (reconstruction change) or a transitional shift (morphology-induced) is present.

10.4.4 GaAs/AlAs (001) Revisited

The different behavior of the GaAs $\beta(2\times4)$ reconstruction upon n- or p-type doping suggests an intimate link between the majority carrier density, carrier type and surface reconstruction. For larger surface Si concentrations, the structure can be switched through several reconstructions at otherwise constant growth conditions by varying only the amount of deposited Si [306]. Since the behavior of the dimer-row kinks is so markedly different for the

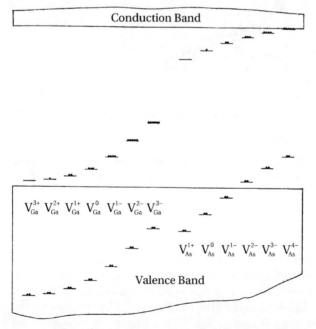

Fig. 10.22. Calculated defect-induced energy levels for undistorted vacancies in GaAs. The symmetry of the upper levels is t_2; for the lower levels it is a_1. The *dots* indicate the occupancy of the defect levels. Adapted from H. Xu and U. Lindefelt [309]

different doping types, we can reverse the argument and investigate whether the presence of kinks can also be associated with the presence of charges. Taking a look at Figs. 7.15b and 7.16b, we note that at both interfaces of the GaAs/AlAs materials system, kinks are present, similar to those present in the case of Si doping. Whereas they form slowly at the normal interface in comparatively small concentrations, they are abundant at the inverted heterointerface. At the same time, deep-level transient spectroscopy (DLTS) measurements [50,310] on GaAs/Al_xGa_{1-x}As heterojunctions show a high concentration of deep-level defects associated with As vacancies and antisites at the inverted interface. These As vacancies are found only above a composition threshold of $x \geq 0.25$. The Al_xGa_{1-x}As layers themselves show a much lower density of intrinsic deep levels over the entire composition range.

The calculated level scheme of the As vacancy in GaAs is shown in Fig. 10.22. Depending on the Fermi-level position, the formation of this vacancy can release up to four electrons. If we assume that the Fermi level remains pinned at midgap at the growing surface, the creation of As vacancies is similar to strong n-type doping.

We can therefore assume that at the normal interface, As vacancies are formed at or very close to the surface as soon as the composition exceeds 25 % of Al. However, during Al_xGa_{1-x}As growth these vacancies are not incorporated in the bulk but segregate at the growth front. At the inverted interface, where there is no Al segregation, the deep-level defects cannot disappear during overgrowth and are incorporated in the bulk. The ionized vacancies then produce high densities of kinks in the immediately overgrown GaAs. The similarity of Figs. 10.18 and 7.16b suggests that the density of the incorporated defects is quite high.

Since these As vacancy defects seem to be intrinsic to the AlAs/GaAs inverted heterointerface, they may be the dominant reason for the observed asymmetry in the transport properties of the two heterointerfaces [45,47, 46] and may mask effects due to the composition variation at the interface. Si doping modifies the RIPS evolution at the normal interface, confirming the correlation of heterointerface formation and electrically active defects. An example is shown in Fig. 10.23. Depending on the Si concentration, the shape of the hump and the initial phase shift at growth initiation are modified, indicating that the segregating Si and/or the incorporated donors modify the electronic structure of the surface, resulting in a different evolution of the surface reconstruction sequence during formation of the heterointerface.

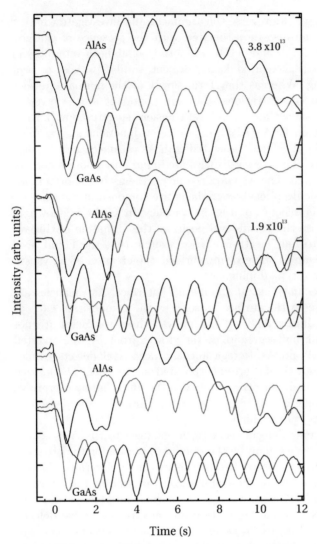

Fig. 10.23. RIPS on the specular spot for two different Si surface concentrations used in the growth sequence of Fig. 10.15 together with the undoped reference (*lowest quadruplet*). The growth and diffraction conditions are identical for all curves

11. Closing Remarks

The aim of this work was to cover applications of RHEED to crystal growth and surface characterization. By looking at RHEED from the 'tool' perspective, we searched for simple and straightforward ways of extracting useful data from the diffraction pattern. In addition to providing information about crystal growth, this approach also yields important insights into the basic processes governing RHEED. At the end of this book, it may be useful to evaluate the current status of RHEED as well as its future potential. How does RHEED compare to other characterization methods? And in what possible directions can RHEED be improved as an in-situ tool?

RHEED is a diffraction method, and as such is limited by the phase problem. Even if the diffraction process were purely kinematical, we would not be able to retrieve the surface structure and morphology directly, but only their autocorrelation function. It is therefore beneficial to complement RHEED with a real-space method such as STM or another scanning microscopy technique to establish a connection between a RHEED phenomenon and the corresponding real-space phenomenon. We used this method for the treatment of domains in Sect. 4.4, and for the characterization of Ga segregation (Sect. 10.1) and Si doping (Sect. 10.4). Since STM, in most cases, is not an in-situ technique, it benefits from the connection to the in-situ situation through RHEED to verify that the structure seen in STM is similar to the one observed by RHEED during epitaxy. RHEED, on the other hand, profits from any real-space data which helps to overcome ambiguities in its interpretation due to the phase problem. Whereas microscopy techniques, including TEM, are often restricted to small sampled areas or volumes, RHEED provides (averaged) information on much larger areas, typically several square millimeters. It is therefore less prone to misinterpretation due to the localized nature of a microscopic phenomenon.

RHEED is a true in-situ tool, and most of its success is based on its relatively simple and straightforward implementation in high-vacuum epitaxy chambers. The cost of a RHEED system is relatively low, especially compared to competing methods like synchrotron X-ray diffraction, and it is therefore a widely used technique. An improvement in the understanding of RHEED therefore immediately benefits a large number of researchers working on many different material systems. The in-situ information that RHEED provides can be used to implement real-time feedback control of the growth process. Compared to other in-situ methods such as ellipsometry, pyrometry and interferometry, RHEED offers the advantage that it does not require access at a steep angle from the front of the substrate, which leaves this space available for deposition equipment. The big advantage of optical methods, however, is their ability to operate under arbitrary growth conditions inside the deposition chamber, especially at high pressures, as long as the medium between the optical port and the substrate is transparent.

RHEED is extremely surface-sensitive, which makes it a prime technique for surface characterization. The experiments presented in the preceding chapters indicate that the dominant diffraction effects take place in the topmost crystal layer. Similar results are found in the literature, especially if surface resonance conditions are applied [311,312]. To obtain similar surface sensitivity with X-rays, high-intensity beams at glancing incidence are required. LEED and He scattering are also very surface-sensitive, but are not as compatible with crystal growth as RHEED and, therefore, are usually not combined with an MBE chamber. On the other hand, the surface sensitivity of RHEED comes at a cost. This is the complicated theoretical treatment because of the strong electron–matter interaction and the resulting multiple scattering effects. Compared to LEED, TEM, X-ray diffraction and even atom scattering, the theoretical understanding of RHEED is less mature. The main reason for this is the strong electron–matter interaction combined with the non-normal truncation of the sample with respect to the beam direction.

If we want to concisely characterize RHEED, we can therefore say that its biggest strength is the combination of surface sensitivity with true in-situ capability. Its most important disadvantage is the complicated theoretical treatment.

It is always difficult to predict future developments in any field of physics. With RHEED, the situation is just the same. Discontinuous developments like the discovery of RHEED oscillations are always possible. The general direction, however, is likely to be both to improve the in-situ capabilities of RHEED and to obtain a practical theoretical model for RHEED on realistic surfaces. This includes, especially, a detailed model of the diffraction mechanism leading to RHEED oscillations. While for a long time the numerical complexity of calculations for systems with a periodicity of more than one surface unit cell put them out of reach, the rapid increase in computer performance has led to several new approaches treating roughness and surface morphology on a larger scale within the framework of a dynamical theory [143,146,147,313]. At the same time, surface reconstruction is being recognized more and more as an important factor in the diffraction process and is incorporated in both large-scale real-space models [313] and one-beam calculations [314]. The explanation of the RHEED oscillation phase in terms of diffraction in the reconstruction layer given in Sect. 9.2 [238] is being adopted in recent publications [313,315]. Progress is also being made in the understanding of the growth mechanism. More detailed atomistic models of growth on a reconstructed surface are being developed [316] that may serve as the basis for more detailed dynamical calculations of RHEED during growth. It is obvious that RHEED theory is an active field of research and significant progress can be expected in the near future. The comparison with and systematic extension of kinematical models may help in the process of developing a comprehensive but simple theory of RHEED. This may allow the application of X-ray techniques like phase determination [317] to RHEED.

On the other hand, we have seen that there is still significant room for improvement in the experimental implementation of RHEED. In particular, the use of measurement on rotating substrates (Chap. 6) may significantly enhance the scope of materials systems and, especially, growth processes that can be examined by RHEED. The ability to measure growth rates in this mode may lead to an increased use of RHEED for growth rate measurements. Energy loss spectroscopy with RHEED (Chap. 8) adds the additional dimension of chemical sensitivity to RHEED, which may be useful for the heteroepitaxy of dissimilar materials and even to detect contamination on the substrate surface [250]. The development of high-resolution RHEED equipment [85,86] as well as new scanning modes (Sect. 6.2) shows new ways to obtain RHEED data, with unprecedented accuracy that may allow more stringent tests of theoretical models.

References

1. K. G. Günther, Z. Naturforsch. **13**A (1958) 1081; W. Hänlein, K.-G. Günther, in *Advances in Vacuum Science and Technology* (Proc. 1st Int. Congr. on Vacuum Techniques, Namur 1958) 727.
2. J. R. Arthur, J. Appl. Phys. **39** (1968) 4032.
3. E. O. Göbel, K. Ploog, Prog. Quant. Electr. **14** (1990) 289.
4. B. V. Shanabrook, J. R. Waterman, J. L. Davis, R. J. Wagner, Appl. Phys. Lett. **61** (1992) 2338.
5. A. J. SpringThorpe, S. J. Ingrey, B. Emmerstorfer, P. Mandeville, W. T. Moore, Appl. Phys. Lett. **50** (1987) 77.
6. F. G. Böbel, H. Möller, B. Hertel, H. Grothe, G. Schraud, St. Schröder, P. Chow, J. Cryst. Growth **150** (1995) 54.
7. S. R. Johnson, T. Tiedje, J. Cryst. Growth **175/176** (1997) 273, and references therein.
8. M. A. Herman, H. Sitter, *Molecular Beam Epitaxy* (Springer Series in Materials Science Vol. 7, 2nd edition, Springer, Berlin, Heidelberg, 1996).
9. L. Esaki, R. Tsu, IBM J. Res. Develop. (Jan. 1970) 61.
10. L. Esaki, R. Tsu, Appl. Phys. Lett. **22** (1973) 562.
11. C. Weisbuch, B. Vinter, *Quantum Semiconductor Structures* (Academic Press, San Diego, 1991).
12. C. Weisbuch, J. Cryst. Growth **127** (1993) 742, and references therein
13. K. Leo, J. Shah, E. O. Göbel, T. C. Damen, S. Schmitt-Rink, W. Schäfer, K. Köhler, Phys. Rev. Lett. **66** (1991) 201.
14. J. Faist, F. Capasso, D. L. Sivco, C. Sirtori, A. L. Hutchinson, A. Y. Cho, Science **264** (1994) 553.
15. L. Pfeiffer, H. L. Störmer, K. W. Baldwin, K. W. West, A. R. Goñi, A. Pinczuk, R. C. Ashoori, M. M. Dignam, W. Wegscheider, J. Cryst. Growth **127** (1993) 849.
16. W. Wegscheider, L. N. Pfeiffer, M. M. Dignam, A. Pinczuk, K. W. West, S. L. McCall, R. Hull, Phys. Rev. Lett. **71** (1993) 4071.
17. M. Ramsteiner, J. Wagner, D. Behr, G. Jungk, L. Däweritz, R. Hey, Appl. Phys. Lett. **64** (1994) 490.
18. M. S. Miller, H. Weman, C. E. Pryor, M. Krishnamurthy, P. M. Petroff, H. Kroemer, J. L. Merz, Phys. Rev. Lett. **68** (1992) 3464.
19. B. Orschel, G. Oelgart, R. Houdré, M. Proctor, F.-K. Reinhart, Appl. Phys. Lett. **62** (1993) 843.
20. T. Schweizer, K. Köhler, P. Ganser, D. J. As, K. H. Bachem, Superlatt. Microstruct. **8** (1990) 179.
21. H. F. Hess, E. Betzig, T. D. Harris, L. N. Pfeiffer, K. W. West, Science **264** (1994) 1740.
22. R. Klann, H. T. Grahn, K. Fujiwara, Phys. Rev. B **51** (1995) 10232.

23. C. A. Warwick, R. F. Kopf, Appl. Phys. Lett. **60** (1992) 386, and references therein.
24. P. M. Young, H. Ehrenreich, Appl. Phys. Lett. **61** (1992) 1069.
25. F. Große, R. Zimmermann, Superlatt. Microstruct. **17** (1995) 439.
26. E. Molinari, S. Baroni, P. Giannozzi, S. de Gironcoli, Phys. Rev. B **45** (1992) 4280.
27. B. Jusserand, F. Mollot, Appl. Phys. Lett. **61** (1992) 423.
28. A. Ourmazd, D. W. Taylor, J. Cunningham, C. W. Tu, Phys. Rev. Lett. **62** (1989) 933.
29. S. Thoma, H. Cerva, Ultramicroscopy **53** (1994) 37.
30. T. Walther, D. Gerthsen, Appl. Phys. A **57** (1993) 393.
31. N. Ikarashi, T. Baba, K. Ishida, Appl. Phys. Lett. **62** (1993) 1632.
32. A. Poudoulec, B. Guenais, C. d'Anterroches, P. Auvray, M. Baudet, A. Regreny, Appl. Phys. Lett. **60** (1992) 2406.
33. J. F. Zheng, X. Liu, N. Newman, E. R. Weber, D. F. Ogletree, M. Salmeron, Phys. Rev. Lett. **72** (1994) 1490.
34. M. B. Johnson, O. Albrektsen, R. M. Feenstra, H. W. M. Salemink, Appl. Phys. Lett. **63** (1993) 2923.
35. M. B. Johnson, U. Maier, H.-P. Meier, H. W. M. Salemink, Appl. Phys. Lett. **63** (1993) 1273.
36. O. Albrektsen, H. P. Meier, D. J. Arent, H. W. M. Salemink, Appl. Phys. Lett. **62** (1993) 2105.
37. E. J. Heller, M. G. Lagally, Appl. Phys. Lett. **60** (1992) 2675.
38. J. Sudijono, M. D. Johnson, M. B. Elowitz, C. W. Snyder, B. G. Orr, Surf. Sci. **280** (1993) 247.
39. M. D. Johnson, J. Sudijono, A. W. Hunt, B. G. Orr, Appl. Phys. Lett. **64** (1994) 484.
40. J. Behrend, M. Wassermeier, W. Braun, P. Krispin, K. H. Ploog, Phys. Rev. B **53** (1996) 9907.
41. J. Behrend, M. Wassermeier, W. Braun, P. Krispin, K. H. Ploog, J. Cryst. Growth **175/176** (1997) 178.
42. C. Gourdon, I. V. Mashkov, P. Lavallard, R. Planel, Phys. Rev. B **57** (1998) 3955.
43. C. T. Foxon, J. J. Harris, D. Hilton, J. Hewett, C. Roberts, Semicond. Sci. Technol. **4** (1989) 582.
44. L. N. Pfeiffer, K. W. West, H. L. Störmer, K. W. Baldwin, Appl. Phys. Lett. **55** (1989) 1888.
45. T. Saku, Y. Hirayama, Y. Horikoshi, Jpn. J. Appl. Phys. **30** (1991) 902.
46. V. Umansky, R. de-Piciotto, M. Heilblum, Appl. Phys. Lett. **71** (1997) 683, and references therein.
47. T. Sajoto, M. Santos, J. J. Heremans, M. Shayegan, M. Heilblum, M. V. Weckwerth, U. Meirav, Appl. Phys. Lett. **54** (1989) 840.
48. H. Shtrikman, A. Soibel, U. Meirav, Appl. Phys. Lett. **72** (1998) 185, and references therein.
49. L. Pfeiffer, E. F. Schubert, K. W. West, C. W. Magee, Appl. Phys. Lett. **58** (1991) 2258.
50. P. Krispin, R. Hey, H. Kostial, J. Appl. Phys. **77** (1995) 5773.
51. J. Massies, F. Turco, A. Saletes, J. P. Contour, J. Cryst. Growth **80** (1987) 307.
52. R. Kohleick, A. Förster, H. Lüth, Phys. Rev. B **48** (1993) 15138.
53. B. Etienne, F. Laruelle, J. Cryst. Growth **127** (1993) 1056.
54. A. J. Springthorpe, P. Mandeville, J. Vac. Sci. Technol. B **6** (1988) 754.

55. A. Lorke, M. Krishnamurty, P. M. Petroff, Mat. Res. Soc. Symp. Proc. **312** (1993) 65.
56. T. Hashizume, Q.-K. Xue, A. Ichimiya, T. Sakurai, Phys. Rev. B **51** (1995) 4200.
57. H. H. Farrell, C. J. Palmstrøm, J. Vac. Sci. Technol. B **8** (1990) 903.
58. W. I. Wang, E. E. Mendez, T. S. Kuan, L. Esaki, Appl. Phys. Lett. **47** (1985) 826.
59. K. Agawa, K. Hirakawa, N. Sakamoto, Y. Hashimoto, T. Ikoma, Appl. Phys. Lett. **65** (1994) 1171.
60. W. Q. Li, P. K. Bhattacharya, S. H. Kwok, R. Merlin, J. Appl. Phys. **72** (1992) 3129.
61. R. Nötzel, N. N. Ledentsov, L. Däweritz, M. Hohenstein, K. Ploog, Phys. Rev. Lett. **67** (1991) 3812.
62. M. Wassermeier, J. Sudijono, M. D. Johnson, K. T. Leung, B. G. Orr, L. Däweritz, K. Ploog, Phys. Rev. B **51** (1995) 14 721.
63. R. Nötzel, K. Ploog, J. Vac. Sci. Technol. A **10** (1992) 617.
64. A. J. Shields, R. Nötzel, M. Cardona, L. Däweritz, K. Ploog, Appl. Phys. Lett. **60** (1992) 2537.
65. O. Brandt, K. Kanamoto, M. Tsugami, T. Isu, N. Tsukada, Appl. Phys. Lett. **67** (1995) 1885.
66. M. Takahashi, P. Vaccaro, K. Fujita, T. Watanabe, Appl. Phys. Lett. **66** (1995) 93.
67. A. A. Kiselev, U. Rössler, Phys. Rev. B **50** (1994) 14 283.
68. Yu. A. Pusep, S. W. da Silva, J. C. Galzerani, D. I. Lubyshev, V. Preobrazhenskii, P. Basmaji, Phys. Rev. B **51** (1995) 5473.
69. J. J. Heremans, M. B. Santos, K. Hirakawa, M. Shayegan, J. Appl. Phys. **76** (1994) 1980.
70. R. Nötzel, J. Temmyo, T. Tamamura, Nature **369** (1994) 131.
71. P. J. Estrup, E. G. McRae, Surf. Sci. **25** (1971) 1.
72. I. Hernández-Calderón, H. Höchst, Phys. Rev. B **27** (1983) 4961.
73. J. M. McCoy, U. Korte, P. A. Maksym, G. Meyer-Ehmsen, Surf. Sci. **261** (1992) 29.
74. A. Ichimiya, K. Kambe, G. Lehmpfuhl, J. Phys. Soc. Japan **49** (1980) 684.
75. L.-M. Peng, J. M. Cowley, N. Yao, Ultramicroscopy **26** (1988) 189.
76. S. W. Bonham, C. P. Flynn, Surf. Sci. **366** (1996) L760.
77. S. L. Dudarev, M. J. Whelan, Int. J. Mod. Phys. **10** (1996) 133.
78. J. L. Beeby, NATO ASI Ser. B **188** (1988) 29.
79. C. Colliex, in *International Tables for Crystallography*, Vol. C (A. J. C. Wilson, ed., Kluwer, Dordrecht, 1992) 337.
80. B. Müller, Fortschr.-Ber. VDI Reihe 9 (VDI, Düsseldorf, 1994).
81. J. J. Quinn, Phys. Rev. **126** (1962) 1453.
82. D. Barlett, C. W. Snyder, B. G. Orr, R. Clarke, Rev. Sci. Instrum. **62** (1991) 1263.
83. Y. Idzerda, J. Vac. Sci. Technol. A **11** (1993) 3138.
84. P. I. Cohen, G. S. Petrich, G. J. Whaley, in *Molecular Beam Epitaxy: Applications to Key Materials* (R. F. C. Farrow, ed., Noyes, Park Ridge, 1995) 669.
85. B. Müller, M. Henzler, Rev. Sci. Instr. **66** (1995) 5232.
86. B. Müller, M. Henzler, Surf. Sci. **389** (1997) 338.
87. W. Tappe, U. Korte, G. Meyer-Ehmsen, Surf. Sci. **388** (1997) 162.
88. H. Nörenberg, L. Däweritz, P. Schützendübe, H.-P. Schönherr, K. Ploog, J. Cryst. Growth **150** (1995) 81.

89. H. Nörenberg, L. Däweritz, P. Schützendübe, K. Ploog, J. Appl. Phys. **81** (1997) 2611.
90. M. Ichikawa, K. Hayakawa, Jpn. J. Appl. Phys. **21** (1982) 145.
91. M. Ichikawa, K. Hayakawa, Jpn. J. Appl. Phys. **21** (1982) 154.
92. M. Hata, A. Watanabe, T. Isu, J. Cryst. Growth **111** (1991) 83.
93. S. Ino, NATO ASI Ser. B **188** (1988) 3.
94. B. F. Lewis, R. Fernandez, A. Madhukar, F. J. Grunthaner, J. Vac. Sci. Technol. B **4** (1986) 560.
95. Ch. Heyn, M. Harsdorff, Appl. Surf. Sci. **100/101** (1996) 494.
96. T. Yokotsuka, M. R. Wilby, D. D. Vvedensky, T. Kawamura, K. Fukutani, S. Ino, Appl. Phys. Lett. **62** (1993) 1673.
97. E. S. Tok, J. H. Neave, F. E. Allegretti, J. Zhang, T. S. Jones, B. A. Joyce, Surf. Sci. **371** (1997) 277.
98. H. Yang, M. Wassermeier, E. Tournié, L. Däweritz, K. Ploog, Surf. Sci. **331** (1995) 479.
99. V. A. Markov, A. I. Nikiforov, O. P. Pchelyakov, J. Cryst. Growth **175/176** (1997) 736.
100. H. Lee, R. Lowe-Webb, W. Yang, P. C. Sercel, Appl. Phys. Lett. **72** (1998) 812.
101. B. Junno, T. Junno, M. S. Miller, L. Samuelson, Appl. Phys. Lett. **72** (1998) 954.
102. H. Raether, in *Encyclopedia of Physics* (Springer, Berlin, Heidelberg, 1957) 443.
103. S. Andrieu, P. Fréchard, Surf. Sci. **360** (1996) 289.
104. L. I. Schiff, *Quantum Mechanics* (3rd edition, McGraw-Hill, New York, 1968).
105. Y. Ma, S. Lordi, P. K. Larsen, J. A. Eades, Surf. Sci. **289** (1993) 47.
106. J. M. McCoy, U. Korte, P. A. Maksym, G. Meyer-Ehmsen, Surf. Sci. **306** (1994) 247.
107. Y. Ma, S. Lordi, P. K. Larsen, J. A. Eades, Surf. Sci. **306** (1994) 252.
108. P. A. Doyle, P. S. Turner, Acta Cryst. A **24** (1968) 390.
109. J. S. Reid, Acta Cryst. A **39** (1983) 1.
110. D. E. Savage, M. Lagally, NATO ASI Ser. B **188** (1988) 475.
111. M. G. Lagally, D. E. Savage, M. C. Tringides, NATO ASI Ser. B **188** (1988) 139.
112. G. Weidemann, J. Griesche, K. Jacobs, J. Cryst. Growth **133** (1993) 75.
113. M. Henzler, NATO ASI Ser. B **188** (1988) 193.
114. P. R. Pukite, P. I. Cohen, S. Batra, NATO ASI Ser. B **188** (1988) 427.
115. K. J. Matysik, Surf. Sci. **38** (1973) 93.
116. L.-M. Peng, M. J. Whelan, Acta Cryst. A **47** (1991) 95.
117. Y. Horio, A. Ichimiya, Surf. Sci. **219** (1989) 128.
118. Y. Horio, A. Ichimiya, S. Kohmoto, H. Nakahara, Surf. Sci. **257** (1991) 167.
119. O. Brandt, H. Yang, K. H. Ploog, Phys. Rev. B **54** (1996) 4432.
120. J. E. Houston, R. L. Park, Surf. Sci. **21** (1970) 209.
121. M. Henzler, Appl. Surf. Sci. **11/12** (1982) 450.
122. P. I. Cohen, G. S. Petrich, P. R. Pukite, G. J. Whaley, A. S. Arrott, Surf. Sci. **216** (1989) 222.
123. C. S. Lent, P. I. Cohen, Surf. Sci. **139** (1984) 121.
124. H.-N. Yang, G.-C. Wang, T.-M. Lu, *Diffraction from Rough Surfaces and Dynamic Growth Fronts* (World Scientific, Singapore, 1993).
125. J. M. Cowley, *Diffraction Physics* (North-Holland, Amsterdam, 1981).
126. L.-M. Peng, S. L. Dudarev, M. J. Whelan, Acta Cryst. A **52** (1996) 471.
127. H. Bethe, Ann. Phys. **87** (1928) 55.
128. K. Shinohara, Inst. Phys. Chem. Res. (Tokyo) **18** (1932) 223.

129. L.-M. Peng, Surf. Sci. **222** (1989) 296.
130. A. Ichimiya, S. Kohmoto, H. Nakahara, Y. Horio, Ultramicroscopy **48** (1993) 425.
131. J. M. Cowley, A. F. Moodie, Acta Cryst. **10** (1957) 609.
132. H. J. Gotsis, P. A. Maksym, Surf. Sci. **385** (1997) 15, and references therein.
133. Y. Ma, S. Lordi, C. P. Flynn, J. A. Eades, Surf. Sci. **302** (1994) 241.
134. Y. Ma, S. Lordi, J. A. Eades, Surf. Sci. **313** (1994) 317.
135. Y. Ma, S. Lordi, J. A. Eades, Phys. Rev. B **49** (1994) 17448.
136. Y. Ma, L. D. Marks, Acta Cryst. A **47** (1991) 707.
137. J. M. McCoy, U. Korte, P. A. Maksym, G. Meyer-Ehmsen, Phys. Rev. B **48** (1993) 4721.
138. M. Witte, G. Meyer-Ehmsen, Surf. Sci. **326** (1995) L499.
139. V. Bressler-Hill, M. Wassermeier, K. Pond, R. Maboudian, G. A. D. Briggs, P. M. Petroff, W. H. Weinberg, J. Vac. Sci. Technol. B **10** (1992) 1881.
140. L. Broekman, R. Leckey, J. Riley, B. Usher, B. Sexton, Surf. Sci. **331–333** (1995) 1115.
141. J. E. Northrup, S. Froyen, Phys. Rev. B **50** (1994) 2015.
142. W. G. Schmidt, F. Bechstedt, Surf. Sci. **360** (1996) L473.
143. Z. L. Wang, Surf. Sci. **366** (1996) 377.
144. U. Korte, J. M. McCoy, P. A. Maksym, G. Meyer-Ehmsen, Phys. Rev. B **54** (1996) 2121.
145. U. Korte, P. A. Maksym, Phys. Rev. Lett. **78** (1997) 2381.
146. G. Meyer-Ehmsen, Surf. Sci. **395** (1998) L189, and references therein.
147. T. Kawamura, Surf. Sci. **351** (1996) 129.
148. J. J. Harris, B. A. Joyce, P. J. Dobson, Surf. Sci. **103** (1981) L90.
149. C. E. C. Wood, Surf. Sci. **108** (1981) L441.
150. J. M. Van Hove, C. S. Lent, P. R. Pukite, P. I. Cohen, J. Vac. Sci. Technol. B **1** (1983) 741.
151. F. Briones, D. Golmayo, L. González, A. Ruiz, J. Cryst. Growth **81** (1987) 19.
152. D. Lee, S. J. Barnett, A. D. Pitt, M. R. Houlton, G. W. Smith, Appl. Surf. Sci. **50** (1991) 428.
153. J. Sudijono, M. D. Johnson, C. W. Snyder, M. B. Elowitz, B. G. Orr, Phys. Rev. Lett. **69** (1992) 2811.
154. T. Shitara, J. Zhang, J. H. Neave, B. A. Joyce, J. Appl. Phys. **71** (1992) 4299.
155. T. Shitara, D. D. Vvedensky, M. R. Wilby, J. Zhang, J. H. Neave, B. A. Joyce, Phys. Rev. B **46** (1992) 6825.
156. J. Hopkins, M. R. Leys, J. Brübach, W. C. van der Vleuten, J. H. Wolter, Appl. Surf. Sci. **84** (1995) 299.
157. J. M. Van Hove, P. R. Pukite, P. I. Cohen, J. Vac. Sci. Technol. B **3** (1985) 563.
158. J. P. A. van der Wagt, K. L. Bacher, G. S. Solomon, J. S. Harris, Jr., J. Vac. Sci. Technol. B **10** (1992) 825.
159. J. E. Cunningham, R. N. Pathak, W. Y. Jan, Appl. Phys. Lett. **68** (1996) 394.
160. G. W. Turner, A. J. Isles, J. Vac. Sci. Technol. B **10** (1992) 1784.
161. J. P. A. van der Wagt, J. S. Harris, Jr., J. Vac. Sci. Technol. B **12** (1993) 1236.
162. G. Grinstein, D. Mukamel, R. Seidin, C. H. Bennett, Phys. Rev. Lett. **70** (1993) 3607.
163. G. S. Petrich, A. M. Dabiran, J. E. Macdonald, P. I. Cohen, J. Vac. Sci. Technol. B **9** (1991) 2150.
164. A. M. Dabiran, S. M. Seutter, P. I. Cohen, Surf. Rev. Lett. **5** (1998) 783.
165. B. A. Joyce, J. Cryst. Growth **99** (1990) 9.

166. L. Däweritz, K. Ploog, Semicond. Sci. Technol. **9** (1994) 123.
167. J. Zhang, J. Neave, P. J. Dobson, B. A. Joyce, Appl. Phys. A **42** (1987) 317.
168. J. Resh, K. D. Jamison, J. Strozier, A. Bensaoula, A. Ignatiev, Phys. Rev. B **40** (1989) 11 799.
169. F. Briones, D. Golmayo, L. González, J. L. De Miguel, Jpn. J. Appl. Phys. **24** (1985) L478.
170. Z. Mitura, M. Strózak, M. Jalochowski, Surf. Sci. Lett. **276** (1992) L15.
171. G. W. Turner, B. A. Nechay, S. J. Eglash, J. Vac. Sci. Technol. B **8** (1990) 283.
172. J. M. van Hove, P. I. Cohen, J. Cryst. Growth **81** (1987) 13.
173. B. Heinrich, M. From, J. F. Cochran, L. X. Liao, Z. Celinski, C. M. Schneider, K. Myrtle, Mat. Res. Soc. Symp. Proc. **313** (1993) 119.
174. J. Massies, N. Grandjean, Phys. Rev. Lett. **71** (1993) 1411.
175. U. May, J. Fassbender, G. Güntherodt, Surf. Sci. **377** (1997) 992.
176. P. I. Cohen, P. R. Pukite, J. M. Van Hove, C. S. Lent, J. Vac. Sci. Technol. A **4** (1986) 1251.
177. J. Wollschläger, Surf. Sci. **383** (1997) 103.
178. L. D. Marks, Y. Ma, Ultramicroscopy **29** (1989) 183.
179. J. H. Neave, B. A. Joyce, P. J. Dobson, N. Norton, Appl. Phys. A **31** (1983) 1.
180. S. Clarke, D. D. Vvedensky, Phys. Rev. Lett. **58** (1987) 2235.
181. P. J. Dobson, B. A. Joyce, J. H. Neave, J. Zhang, J. Cryst. Growth **81** (1987) 1.
182. B. A. Joyce, J. H. Neave, J. Zhang, P. J. Dobson, NATO ASI Ser. B **188** (1988) 397.
183. D. M. Holmes, J. L. Sudijono, C. F. McConville, T. S. Jones, B. A. Joyce, Surf. Sci. **370** (1997) L173.
184. S. L. Dudarev, D. D. Vvedensky, M. J. Whelan, Phys. Rev. B **50** (1994) 14525.
185. S. L. Dudarev, D. D. Vvedensky, M. J. Whelan, Surf. Sci. **324** (1995) L355.
186. B. G. Orr, M. D. Johnson, C. Orme, J. Sudijono, A. W. Hunt, Solid State Electron. **37** (1994) 1057.
187. Ch. Heyn, T. Franke, R. Anton, M. Harsdorff, Phys. Rev. B **56** (1997) 13 483.
188. L.-M. Peng, M. J. Whelan, Surf. Sci. **238** (1990) L446.
189. L.-M. Peng, M. J. Whelan, Proc. R. Soc. Lond. A **432** (1991) 195.
190. L.-M. Peng, M. J. Whelan, Proc. R. Soc. Lond. A **435** (1991) 257.
191. L.-M. Peng, M. J. Whelan, Proc. R. Soc. Lond. A **435** (1991) 269.
192. Y. Horio, A. Ichimiya, Surf. Sci. **298** (1993) 261.
193. R. Nötzel, K. H. Ploog, Int. J. Mod. Phys. B **7** (1993) 2743.
194. L. Däweritz, R. Nötzel, K. H. Ploog, SPIE Proc. **2141** (1994) 114.
195. G.-X. Qian, R. M. Martin, D. J. Chadi, Phys. Rev. B **38** (1988) 7649.
196. W. Braun, O. Brandt, M. Wassermeier, L. Däweritz, K. Ploog, Appl. Surf. Sci. **104/105** (1996) 35.
197. C. Setzer, J. Platen, P. Geng, W. Ranke, K. Jacobi, Surf. Sci. **377–379** (1997) 125.
198. P. R. Watson, M. A. Van Hove, K. Hermann, *Atlas of Surface Structures* (J. Phys. Chem. Ref. Data Monograph No. 5, American Chemical Society, Washington, DC, 1994).
199. Y. Hsu, W. I. Wang, T. S. Kuan, Phys. Rev. B **50** (1994) 4973.
200. D. Lüerßen, A. Dinger, H. Kalt, W. Braun, R. Nötzel, K. H. Ploog, J. Tümmler, J. Geurts, Phys. Rev. B **57** (1998) 1631.
201. D. I. Lubyshev, M. Micovic, D. L. Miller, I. Chizhov, R. F. Willis, J. Vac. Sci. Technol. B **16** (1998) 1339.
202. T. Hashizume, Q. K. Xue, J. Zhou, A. Ichimiya, T. Sakurai, Phys. Rev. Lett. **73** (1994) 2208.

203. W. Braun, PhD thesis (Berlin, 1996); HTML version available online at http://asumbe.eas.asu.edu/wolfgang/welcome.htm.
204. J. Griesche, J. Cryst. Growth **149** (1995) 141.
205. A. Weickenmeier, H. Kohl, Acta Cryst. A **47** (1991) 590.
206. J. Behrend, M. Wassermeier, L. Däweritz, K. H. Ploog, Surf. Sci. **342** (1995) 63.
207. H. Oigawa, M. Wassermeier, J. Behrend, L. Däweritz, K. H. Ploog, Surf. Sci. **376** (1997) 185.
208. K. Fujiwara, J. Phys. Soc. Japan **12** (1957) 7.
209. M. R. Fahy, M. J. Ashwin, J. J. Harris, R. C. Newman, B. A. Joyce, Appl. Phys. Lett. **61** (1992) 1805.
210. J. G. Belk, C. F. McConville, J. L. Sudijono, T. S. Jones, B. A. Joyce, Surf. Sci. **387** (1997) 213.
211. L. Däweritz, K. Stahrenberg, P. Schützendübe, J.-T. Zettler, W. Richter, K. H. Ploog, J. Cryst. Growth **175/176** (1997) 310.
212. J. M. Moison, C. Guille, M. Bensoussan, Phys. Rev. Lett. **58** (1987) 2555.
213. A. Sugawara, K. Fujieda, N. Otsuka, Surf. Sci. **394** (1997) L174.
214. M. Wassermeier, S. Kellermann, J. Behrend, L. Däweritz, K. Ploog, Surf. Sci. **414** (1998) 298.
215. M. Wassermeier, A. Yamada, H. Yang, O. Brandt, J. Behrend, K. H. Ploog, Surf. Sci. **385** (1997) 178.
216. S. Pflanz, W. Moritz, Acta Cryst. A **48** (1992) 716, and references therein.
217. S. Pflanz, H. L. Meyerheim, W. Moritz, I. K. Robinson, H. Hoernis, E. H. Conrad, Phys. Rev. B **52** (1995) 2914.
218. M. Sauvage-Simkin, Y. Garreau, R. Pinchaux, M. B. Véron, J. P. Landesman, J. Nagle, Phys. Rev. Lett. **75** (1995) 3485.
219. S. Kikuchi, Jpn. J. Phys. **5** (1928) 83.
220. Y. Kainuma, Acta Cryst. **8** (1955) 247, and references therein.
221. M. Gajdardziska-Josifovska, J. M. Cowley, Acta Cryst. A **47** (1991) 74, and references therein.
222. P. J. Dobson, NATO ASI Ser. B **191** (1988) 159.
223. N. Yao, J. M. Cowley, Ultramicroscopy **31** (1989) 149.
224. L.-M. Peng, J. K. Gjønnes, Acta Cryst. A **45** (1989) 699.
225. Z. L. Wang, J. M. Cowley, Ultramicroscopy **26** (1988) 233.
226. J. E. Griffith, D. A. Grigg, J. Appl. Phys. **74** (1993) R83.
227. P. B. Hirsch, A. Howie, R. B. Nicholson, D. W. Pashley, M. J. Whelan, *Electron Microscopy of Thin Crystals* (Butterworth, London, 1971) 496.
228. Z. Mitura, P. A. Maksym, Phys. Rev. Lett. **70** (1993) 2904.
229. Z. Mitura, P. Mazurek, K. Paprocki, P. Mikołajczak, Appl. Phys. A **60** (1995) 227.
230. Z. Mitura, P. Mazurek, K. Paprocki, P. Mikołajczak, J. L. Beeby, Phys. Rev. B **53** (1996) 10 200.
231. Z. Mitura, J. L. Beeby, J. Phys. Condens. Matter **8** (1996) 8717.
232. R. F. Kromann, R. N. Bicknell-Tassius, A. S. Brown, J. F. Dorsey, K. Lee, G. May, J. Cryst. Growth **175/176** (1997) 334.
233. W. Braun, H. Möller, Y.-H. Zhang, J. Vac. Sci. Technol. B **16** (1998) 1507.
234. G. Meyer-Ehmsen, NATO ASI Ser. B **188** (1988) 99.
235. J. H. Neave, B. A. Joyce, P. J. Dobson, Appl. Phys. A **34** (1984) 179.
236. Y. Horio, A. Ichimiya, Surf. Sci. **219** (1993) 128.
237. J. S. Resh, K. D. Jamison, J. Strozier, A. Bensaoula, A. Ignatiev, Vacuum **41** (1990) 1052.
238. W. Braun, L. Däweritz, K. H. Ploog, Phys. Rev. Lett. **80** (1998) 4935.
239. W. Braun, L. Däweritz, K. H. Ploog, J. Vac. Sci. Technol. B **16** (1998) 2404.

240. D. A. Collins, G. O. Papa, T. C. McGill, J. Vac. Sci. Technol. B **13** (1995) 1953.
241. Y. Ma, S. Lordi, J. A. Eades, Proc. Mat. Res. Soc. **399** (1996) 3.
242. A. M. Dabiran, P. I. Cohen, J. Cryst. Growth **150** (1995) 23.
243. H. Morcoç, T. J. Drummond, W. Kopp, R. Fischer, J. Electrochem. Soc. **129** (1982) 824.
244. P. B. Hirsch, A. Howie, R. B. Nicholson, D. W. Pashley, M. J. Whelan, *Electron Microscopy of Thin Crystals* (Butterworth, London, 1971) 491.
245. P. M. Petroff, A. C. Gossard, W. Wiegmann, A. Savage, J. Cryst. Growth **44** (1978) 5.
246. M. D. Pashley, K. W. Haberern, J. M. Gaines, Appl. Phys. Lett. **58** (1991) 406.
247. H. A. Atwater, C. C. Ahn, Appl. Phys. Lett. **58** (1991) 269.
248. C. C. Ahn, S. Nikzad, H. A. Atwater, Mat. Res. Soc. Symp. Proc. **208** (1991) 251.
249. S. Nikzad, C. C. Ahn, H. A. Atwater, J. Vac. Sci. Technol. B **10** (1992) 762.
250. S. Nikzad, S. S. Wong, C. C. Ahn, A. L. Smith, H. A. Atwater, Appl. Phys. Lett. **63** (1993) 1414.
251. H. A. Atwater, S. S. Wong, C. C. Ahn, S. Nikzad, H. N. Frase, Surf. Sci. **298** (1993) 273.
252. M. M. Disko, C. C. Ahn, B. Fultz (eds.), *Transmission Electron Energy Loss Spectrometry in Materials Science* (The Minerals, Metals & Materials Society, Warrendale, 1992).
253. E. J. Scheibner, L. N. Tharp, Surf. Sci. **8** (1967) 247.
254. H. Boersch, Z. Phys. **134** (1953) 156.
255. Y. Horio, Y. Hashimoto, K. Shiba, A. Ichimiya, Jpn. J. Appl. Phys. **34** (1995) 5869.
256. Y. Horio, Y. Hashimoto, A. Ichimiya, Appl. Surf. Sci. **100/101** (1996) 292.
257. J. M. Cowley, J. L. Albain, G. G. Hembree, P. E. Højlund-Nielsen, F. A. Koch, J. D. Landry, H. Shuman, Rev. Sci. Instr. **46** (1975) 826
258. C. J. Powell, Phys. Rev. **175** (1968) 972
259. A. L. Bleloch, A. Howie, R. H. Milne, M. G. Walls, NATO ASI Ser. B **188** (1988) 77.
260. C. S. Lent, P. I. Cohen, Phys. Rev. B **33** (1986) 8329.
261. M. Tanaka, H. Sakaki, Superlatt. Microstruct. **4** (1988) 237.
262. H. Raether, *Excitation of Plasmons and Interband Transitions by Electrons* (Springer Tracts in Modern Physics, Vol. 88, Springer, Berlin, Heidelberg, 1980) 116.
263. W. Braun, L. Däweritz, K. H. Ploog, Surf. Sci. **399** (1998) 234.
264. J. Falta, R. M. Tromp, M. Copel, G. D. Pettit, P. D. Kirchner, Phys. Rev. Lett. **69** (1992) 3068.
265. D. K. Biegelsen, R. D. Bringans, J. E. Northrup, L.-E. Swartz, Phys. Rev. B **41** (1990) 5701.
266. H. Nörenberg, N. Koguchi, Surf. Sci. **296** (1993) 199.
267. G. Lehmpfuhl, A. Ichimiya, H. Nakahara, Surf. Sci. **245** (1991) L159.
268. Y. Horio, A. Ichimiya, Jpn. J. Appl. Phys. **33** (1994) L377.
269. L. Däweritz, J. Cryst. Growth **127** (1993) 949.
270. B. A. Joyce, P. J. Dobson, J. H. Neave, Surf. Sci. **178** (1986) 110.
271. M. D. Johnson, C. Orme, A. W. Hunt, D. Graff, J. Sudijono, L. M. Sander, B. G. Orr, Phys. Rev. Lett. **72** (1994) 116.
272. G. S. Spencer, J. Menéndez, L. N. Pfeiffer, K. W. West, Phys. Rev. B **52** (1995) 8205.
273. V. Thierry-Mieg, F. Laruelle, B. Etienne, J. Cryst. Growth **127** (1993) 1022.

274. K. Ohta, Surf. Sci. **298** (1993) 415.
275. W. Braun, A. Trampert, L. Däweritz, K. H. Ploog, Phys. Rev. B **55** (1997) 1689.
276. J. M. Moison, C. Guille, F. Houzay, F. Barthe, M. Van Rompay, Phys. Rev. B **40** (1989) 6149.
277. D. McLean, *Grain Boundaries in Metals* (Oxford University Press, Oxford, 1957) 118.
278. E. D. Hondros, in *Grain Boundary Structure and Properties* (G. A. Chadwick, D. A. Smith, eds., Academic Press, London, 1976) 265–299.
279. J. Bénard, Y. Berthier, F. Delamare, E. Hondros, M. Huber, P. Marcus, A. Masson, J. Oudar, G. E. Rhead, *Adsorption on Metal Surfaces* (Elsevier, Amsterdam, Oxford, New York, 1983) 245–270.
280. J. P. Landesman, J. Nagle, J. C. Garcia, C. Mottet, M. Larive, J. Massies, G. Jezequel, P. Bois, in *Semiconductor Interfaces at the Sub-Nanometer Scale*, NATO ASI Ser. E **243** (H. W. M. Salemink, M. D. Pashley, eds.,Kluwer, Dordrecht, Boston, London 1993) 105–113.
281. S. Fukatsu, K. Fujita, H. Yaguchi, Y. Shiraki, R. Ito, Appl. Phys Lett. **59** (1991) 2103.
282. K. Fujita, S. Fukatsu, H. Yaguchi, Y. Shiraki, R. Ito, Appl. Phys. Lett. **59** (1991) 2240.
283. S. V. Ivanov, P. S. Kop'ev, N. N. Ledentsov, J. Cryst. Growth **104** (1990) 345.
284. S. V. Ivanov, P. S. Kop'ev, N. N. Ledentsov, J. Cryst. Growth **111** (1991) 151.
285. P. F. Fewster, N. L. Andrew, C. J. Curling, Semicond. Sci. Technol. **6** (1991) 5.
286. M. Ilg, K. H. Ploog, Phys. Rev. B **48** (1993) 11 512.
287. J. J. Harris, D. E. Ashenford, C. T. Foxon, P. J. Dobson, B. A. Joyce, Appl. Phys. A **33** (1984) 87.
288. K. Ploog, A. Fischer, J. Vac. Sci. Technol. **15** (1978) 255.
289. T. J. de Lyon, J. M. Woodall, M. S. Goorsky, P. D. Kirchner, Appl. Phys. Lett. **56** (1990) 1040.
290. R. J. Malik, J. Nagle, M. Micovic, T. Harris, R. W. Ryan, L. C. Hopkins, J. Vac. Sci. Technol. B **10** (1992) 850.
291. C. Giannini, A. Fischer, C. Lange, K. Ploog, L. Tapfer, Appl. Phys. Lett. **61** (1992) 183.
292. G. Burns, *Solid State Physics* (Academic Press, Boston, 1990) 315.
293. M. D. Pashley, K. W. Haberern, R. M. Feenstra, P. D. Kirchner, Phys. Rev. B **48** (1993) 4612.
294. M. D. Pashley, K. W. Haberern, Phys. Rev. Lett. **67** (1991) 2697.
295. M. D. Pashley, K. W. Haberern, R. M. Feenstra, J. Vac. Sci. Technol. B **10** (1992) 1874.
296. K. Ploog, J. Cryst. Growth **81** (1987) 304.
297. E. F. Schubert, J. M. Kuo, R. F. Kopf, A. S. Jordan, H. S. Luftman, L. C. Hopkins, Phys. Rev. B **42** (1990) 1364.
298. M. T. Asom, G. Livescu, M. Geva, V. Swaminathan, L. C. Luther, R. E. Leibenguth, V. D. Mattera, E. F. Schubert, J. M. Kuo, R. Kopf, J. Cryst. Growth **111** (1991) 246.
299. J. J. Harris, R. B. Beall, J. B. Clegg, C. T. Foxon, S. J. Battersby, D. E. Lacklison, G. Duggan, C. M. Hellon, J. Cryst. Growth **95** (1989) 257.
300. M. J. Ashwin, M. Fahy, J. J. Harris, R. C. Newman, D. A. Sansom, R. Addinall, D. S. McPhail, V. K. M. Sharma, J. Appl. Phys. **73** (1992) 633.
301. O. Brandt, G. E. Crook, K. Ploog, J. Wagner, M. Maier, Appl. Phys. Lett. **59** (1991) 2730.

302. L. Däweritz, H. Kostial, M. Ramsteiner, R. Klann, P. Schützendübe, K. Stahrenberg, J. Behrend, R. Hey, M. Maier, K. Ploog, Phys. Stat. Sol. **194** (1996) 127.

303. B. Chen, Q.-M. Zhang, J. Bernholc, Phys. Rev. B **49** (1994) 2985.

304. G. A. Baraff, M. Schlüter, Phys. Rev. Lett. **55** (1985) 1327.

305. J. Wang, T. A. Arias, J. D. Joannopoulos, G. W. Turner, O. L. Alerhand, Phys. Rev. B **47** (1993) 10 326.

306. M. Wassermeier, J. Behrend, L. Däweritz, K. Ploog, Phys. Rev. B **52** (1995) R2269.

307. K. Kanisawa, H. Yamaguchi, Y. Horikoshi, J. Cryst. Growth **175/176** (1997) 304.

308. O. Brandt, G. Crook, K. Ploog, R. Bierwolf, M. Hohenstein, M. Maier, J. Wagner, Jpn. J. Appl. Phys. **32** (1993) L24.

309. H. Xu, U. Lindefelt, Phys. Rev. B **41** (1990) 5979.

310. P. Krispin, R. Hey, H. Kostial, K. H. Ploog, J. Appl. Phys. **83** (1998) 1496.

311. S. W. Bonham, C. P. Flynn, Phys. Rev. B **57** (1998) 4099.

312. Y. Horio, Phys. Rev. B **57** (1998) 4736.

313. M. Itoh, Phys. Rev. B **58** (1998) 6716.

314. E. A. Khramtsova, Phys. Rev. B **57** (1998) 10 049.

315. Z. Mitura, S. L. Dudarev, M. J. Whelan, Phys. Rev. B **57** (1998) 6309.

316. M. Itoh, G. R. Bell, A. R. Avery, T. S. Jones, B. A. Joyce, D. D. Vvedensky, Phys. Rev. Lett. **81** (1998) 633.

317. Q. Shen, Phys. Rev. Lett. **80** (1998) 3268.

Index

Color Plates

This section contains color reproductions of Figs. 7.12, 7.13, 7.14, 7.18, 10.10, 10.13 and 10.18. Please note that, due to space limitations, Fig. 10.18 has been placed between Figs. 7.12 and 7.13.

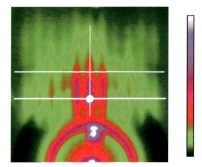

Fig. 7.12. (see original on p. 121)

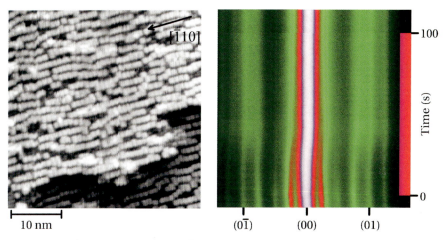

Fig. 10.18. (see original on p. 188)

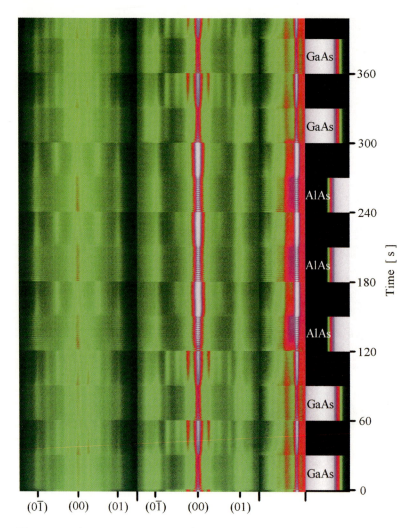

Fig. 7.13. (see original on p. 122)

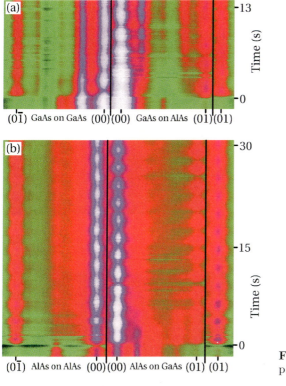

(0$\bar{1}$) GaAs on GaAs (00)(00) GaAs on AlAs (01)(01)

(0$\bar{1}$) AlAs on AlAs (00)(00) AlAs on GaAs (01) (01)

Fig. 7.14. (see original on p. 123)

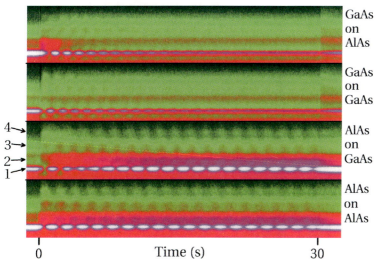

GaAs on AlAs

GaAs on GaAs

4→
3→
2→
1→

AlAs on GaAs

AlAs on AlAs

0   Time (s)   30

Fig. 7.18. (see original on p. 127)

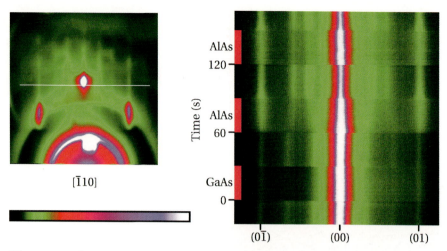

Fig. 10.10. (see original on p. 180)

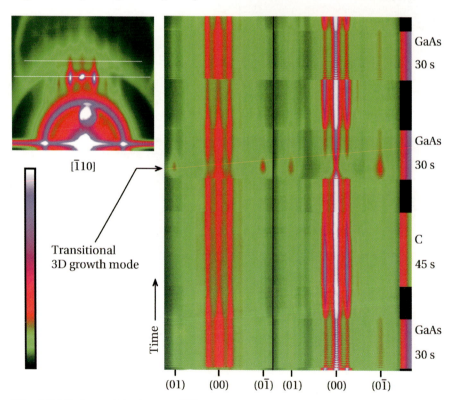

Fig. 10.13. (see original on p. 183)

Springer Tracts in Modern Physics

Springer
and the
environment

At Springer we firmly believe that an international science publisher has a special obligation to the environment, and our corporate policies consistently reflect this conviction.

We also expect our business partners – paper mills, printers, packaging manufacturers, etc. – to commit themselves to using materials and production processes that do not harm the environment. The paper in this book is made from low- or no-chlorine pulp and is acid free, in conformance with international standards for paper permanency.